THE FIFTH SEASON

THE FIFTH SEASON

THE CHICAGO TREE PROJECT

Margot McMahon

Tess Landon

MargotMcMahon.com

HUMMINGBIRD BOOKS

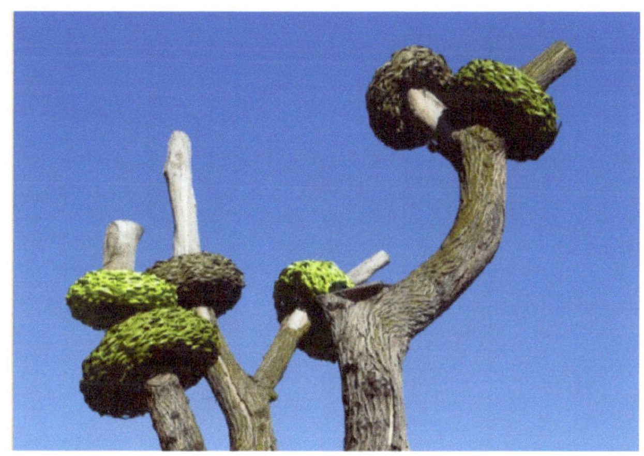

Karen Gubitz

CHICAGO TREE PROJECT
The Leaving Tree

Margot McMahon

Photographs
Tess Landon and Margot McMahon

@2021 Margot McMahon

Margot McMahon received the
2019 Mate E. Palmer First Place Service Award
2020 Mate E. Palmer First Place Book Award
Illinois Women's Press Association
for the Chicago Tree Project

Margot McMahon Statement

Chicago Tree Project celebrated its fifth season with fifty sculpted trees in over thirty miles of Chicago Parks. Neighbors embrace the public art statements while mourning the loss of a mature canopy. Sculpting dead trees gives them a fifth season harboring food and shelter for wildlife. From grave to cradle, the Chicago Tree Project will teach young people how to care for the hundreds of thousands of saplings planted to replace our lost canopy. Sculpting grand condemned trees, caring for replacement young plantlets, and learning that carbon is choking both young and old is our mission. Warming temperatures and more volatile weather stresses trees making it more difficult for saplings to thrive. We imagined teens, planting a sapling in another neighborhood. They nurture another group's sprout near their home. Annually, they picnic at one sapling, to compare notes: watering, mulching, and how raising an Oak is like raising a Maple. Teens learn to care for young urban orchards, their future oxygen source. Sapling-Pals, think of their tree elsewhere and care for another group's sprig at home.

Visit the website: chicagotreeproject.org for links to artists' information

Imprint

Any brand names and product names mentioned in this book are subject to trademark, brand or patent protection and are trademarks or registered trademarks of their respective holders. The use of brand names, product names, common names, trade names, production descriptions etc. even without a particular marking in this work is in no way to be construed to mean that such names may be regarded as unrestricted in respect of trademark and brand protection legislation and could thus be used by anyone.

Cover image: *Flock* Margot McMahon
Publisher:
Hummingbird Books
310 S. Humphrey Ave,
Oak Park, IL 60302
Printed at
ISBN: 978-1-0879-0172-5
Copyright Margot McMahon 2021

Acknowledgments

The Chicago Tree Project has been made possible by the collaboration of Chicago Sculpture International, the Chicago Park District, the City of Chicago, and the creativity of artists. For ensuring the continuing nature of making artistic statements, Michael Dimitroff, Carolyn Bendel, Tess Landon, and Christine Perri are gratefully acknowledged. Thanks to Daniel J. Burke for editing.

Margot McMahonhas gratefully received the
2019 Mate E. Palmer First Place Service Award
2020 First Place Book Award
The Illinois Women's Press Association
for her contributions to the
Chicago Tree Project.

Margot McMahon Statement

The Chicago Tree Project celebrated its fifth season with fifty sculpted trees in over thirty miles of Chicago Parks. Neighbors embrace the public art statements while mourning the loss of mature tree canopies. Sculpting dead trees gives them a fifth season by harboring food and shelter for wildlife. From grave to cradle, the Chicago Tree Project will teach young tree-tenders of all ages how to care for the millions of saplings being planted to replace our lost canopy. Sculpting grand condemned trees with artistic statements, caring for replacement young plantlets, and showing that carbon is choking both young and old is our mission. Warming temperature and more volatile weather stresses trees making it more difficult for saplings to thrive.

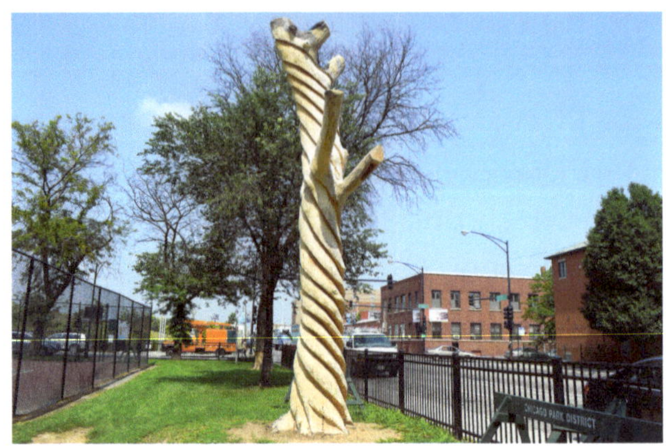

Introduction

Nature has solved the problem of how to catch in flight, light streaming to the earth, and store this most elusive of all powers in rigid form. Dr. Julius Robert Mayer, 1880s.

Why Sculpt Dead Trees?

The Chicago Tree Project, an initiative to increase awareness of the dying population of various species of trees due to climate change in Chicago's urban parks, celebrated its fifth season. Approximately fifty international sculptors have transformed condemned elm, ash, and locust trees into artistic statements along thirty miles of Chicago parks. Artists through Chicago Sculpture International have teamed up with the Chicago Park District to select trees offering an unusual canvas for artists to create their concepts. Their activism incites the art, and their art makes the activism direct and palatable. The Chicago Tree project applauds the fearless and honest activist artists who symbolize hard truths through clarity of transforming nature into art through mastery. The mastery of making art is evident in Ron Gard's whittled-with-a-chainsaw *Dying to Survive*, Anthony Heinz May's branches made into dynamic, cubed lumber *Natura Non Confundenda* and *Progress>(Product),* and Christine Perri's architectural referencing with *De-Constructed Monument*.

The Chicago Tree Project gives urban trees a fifth season by sculpting them. Artists invent concepts for their sculpted trees to show nature in an urban environment. Mia Capodilupo's *Burst,*

Allison Svoboda's *Ghost Tree* and Bryan Northrup's *Message in a Bottle,* wrap industrial materials around condemned trees to inspire contemplation through the aesthetic imagery of industry choking nature. Janet Austin's bronze Ash Borer larvae interrupts, with a maze path, the nourishing cambium trunk layer in her sculpted tree, *The A"maze"ing Larvae of the Emerald Ash Borer,* Marc Schneider's *Invasive Species* refers to a car-size threatening Asian Carp stick-assembled birdhouse referencing *Moby Dick* in its presence. Spirals are found in J. Taylor Wallace's *Aspire,* Anne Alexander's *Solspirals* and Phillip Shore's *Eternal Connections* expressing pleasing twisting life forms and ancient universal symbols.

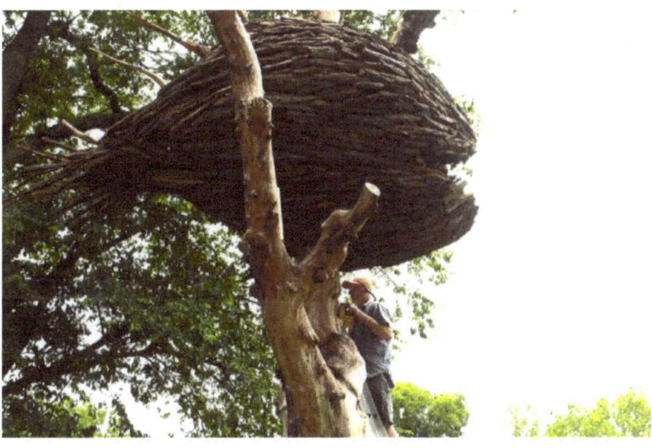

These sculptures in trees draw attention to our live-condemned trees and pays tribute to their decades or centuries of life. Chicago sculptors are recreating trees for a variety of reasons. Some want to draw attention to the large number of trees dying, some want to say their work is part of the urban natural environment, some want to create a symbol of new hope, shelter and food for animal life. Some want to remind others to reduce their carbon footprint by easing your foot off the peddle. All artists, perched in these branches, are making a statement in this unique medium.

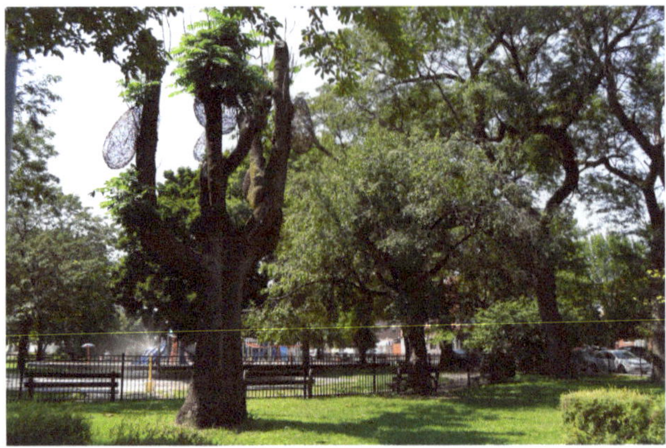

The sculpted trees are selected for parks throughout Chicago, creating a ripple effect for artists' expressions through neighborhoods, some of which are art deserts. These sculptures are a few reminders of the importance of these trees in our heavily populated areas.

Figurative sculptures reference human responsibility of excessive carbon emissions in Kara James peaceful *Lead from the Heart* and *Save the Birds* as well as Margot McMahon's fanciful skirmishing between *Queen* and *Knight* chess pieces of *Checkmate* in Lincoln Park. Steven Gaeth's *Recompense* genuflects to show the diversity and unity of spirituality. Human hands, nurturing nature, are symbolized in Indira and Karl Johnson's *SOS* patterns and Carrie Fischer's *Helping Hand*. Artists consider the plight of urban animals through JR Cadawas' *Dog Tree*, Jim Long's *Eagle, Heron* and *Coyote*, Tracy Ostmann Haschke's reptilian *Fraxinus Raptor*, Margot McMahon's variety of wild and domestic avian species in Columbus Park's *Flock* and Hale Park's *Perch•Preen* as well as Sandra Bacon and John Hatlestad's *Fancy Birds* in Lincoln Park. Creating places to shelter animals, insects and birds are found in Vivian Visser's woven willow *Hive, Cocoons, Ledge Fungus*, Karen Gubitz's master basketry in *Gaggle, Second Chances*, Samantha Rausch's vibrant, threaded *Endless*, Peter Krsko's prefabricated inviting modules of *Para Z* and Nicolette Ross' etched *Tapestry*.

Titles are clues to the thoughtful activism and caring concerns of Chicago Tree Project artists. Gary Lehman's *Shards of the Rainbow*, Gail Simpson and Aristotle Georgiades' *Conundrum*,

Connie Noyes' *The Sound of Light on Ash* and Nick Goettling's *Decades* are aesthetic images, with descriptive titles strung together into prose, to make an expressive tribute to trees, moral fiber and poetic reverence for keeping dying trunks as a part of our lives.

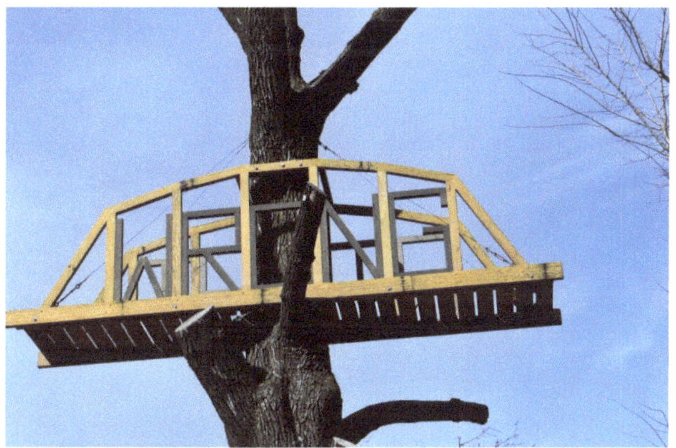

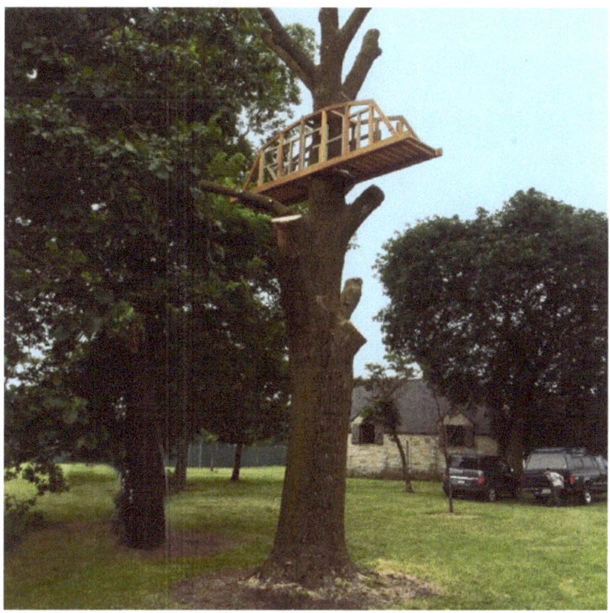

These sculptures are a few reminders of the importance of trees in our heavily populated areas. Chicago is proud to tell the world of our artists and visiting artists who have represented new growth through the delicate blossoms of Thom Cicchelli's *Arbor Renewal* and Cheryl William's *Tree Hugger*. Chicago is proud to bring artistic statements to neighborhoods, previously without art, to elevate life, think of concepts beyond the daily schedule and allow more wonder. Chicago is proud of the Chicago Tree Project and its artists.

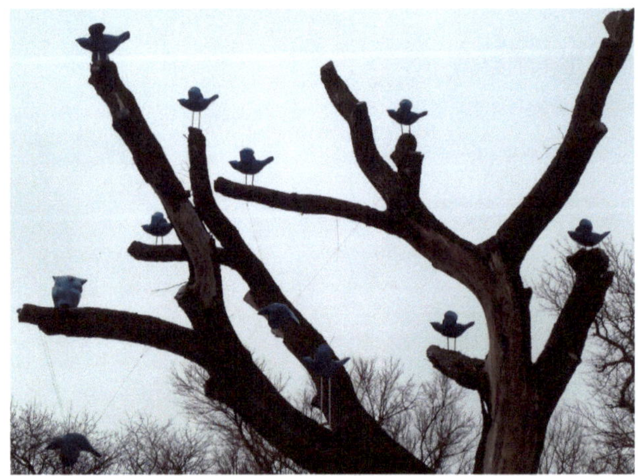

Street Cred

Canopy of light green leaves
Clean the soiled city breeze
Monopoly of sun-showered young tree
Tripoli, Chicago, New York city's

Remedy condemned and dying trees.
This sapling, our transplanted Street Kid,
Makes its lonesome way
Standing tall along parkways among
Cousins, neighbors, a few new friends.

On its own it must contend.
Not a brother, sister, mother, father
To web its roots, share its fodder
Streets kids die young
The demise of carbon consumption,
Street kids untethered, oxygen exhaling.

Extending tendrils in search of kin
Street kids gulp sun with abandon
Stretched tall fast, no shade thrown
Gobble carbon from its hood
Spew oxygen as it should
Tired young without shared trees
Screams withered of lonesome plea

Street Kids, don't grow with abandon
Slow your bravado, calm your exhaustion
Slow your mein that spent your cred

Store your reserve, elude the dread.

-----Margot McMahon

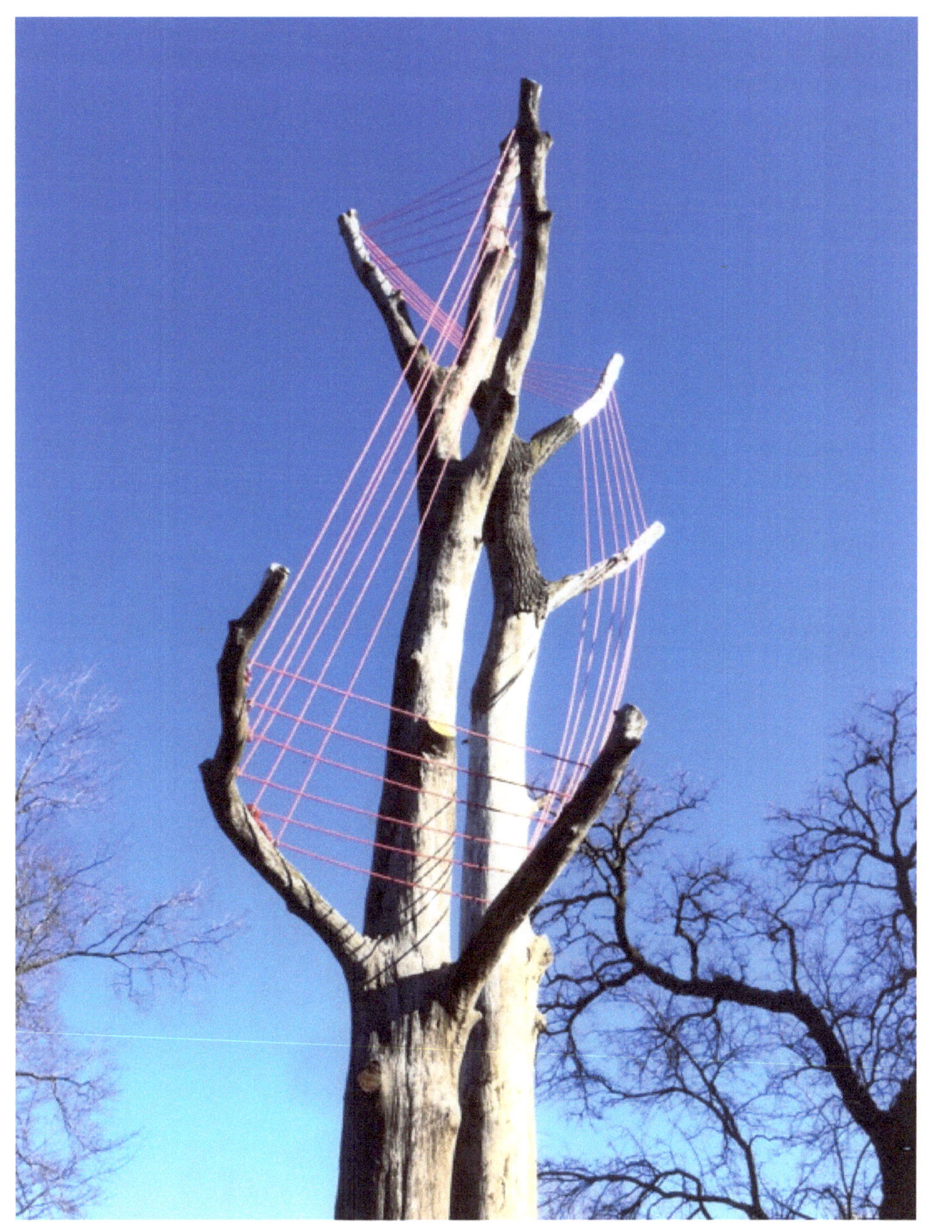

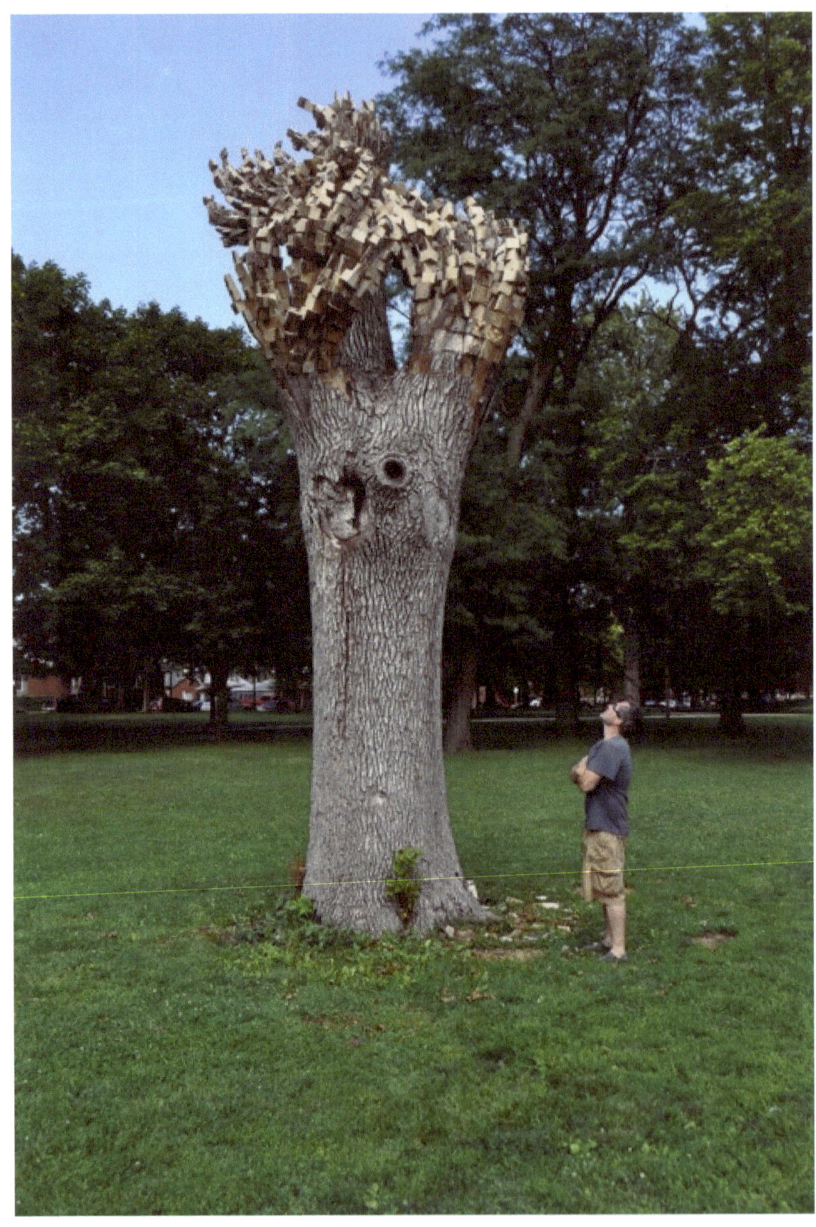

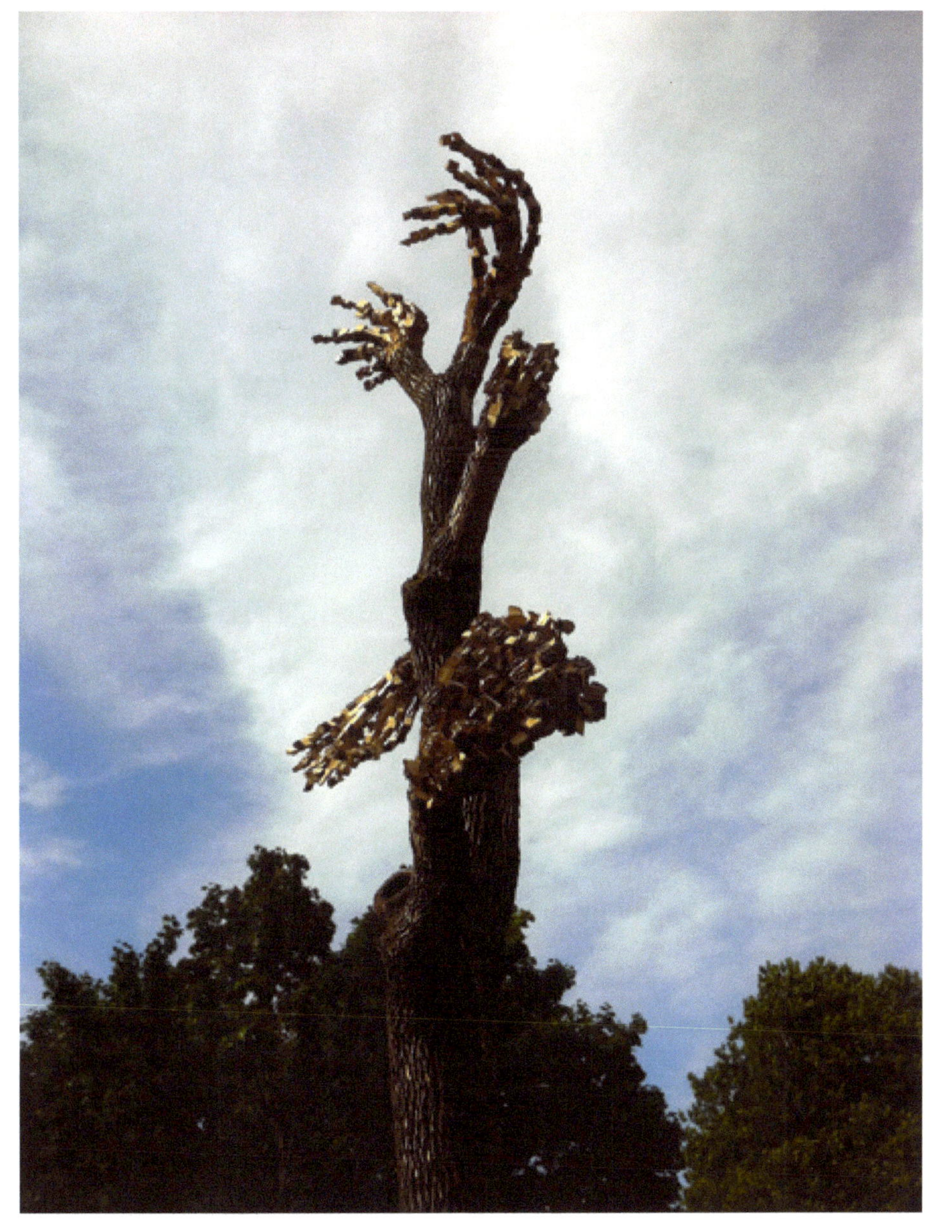

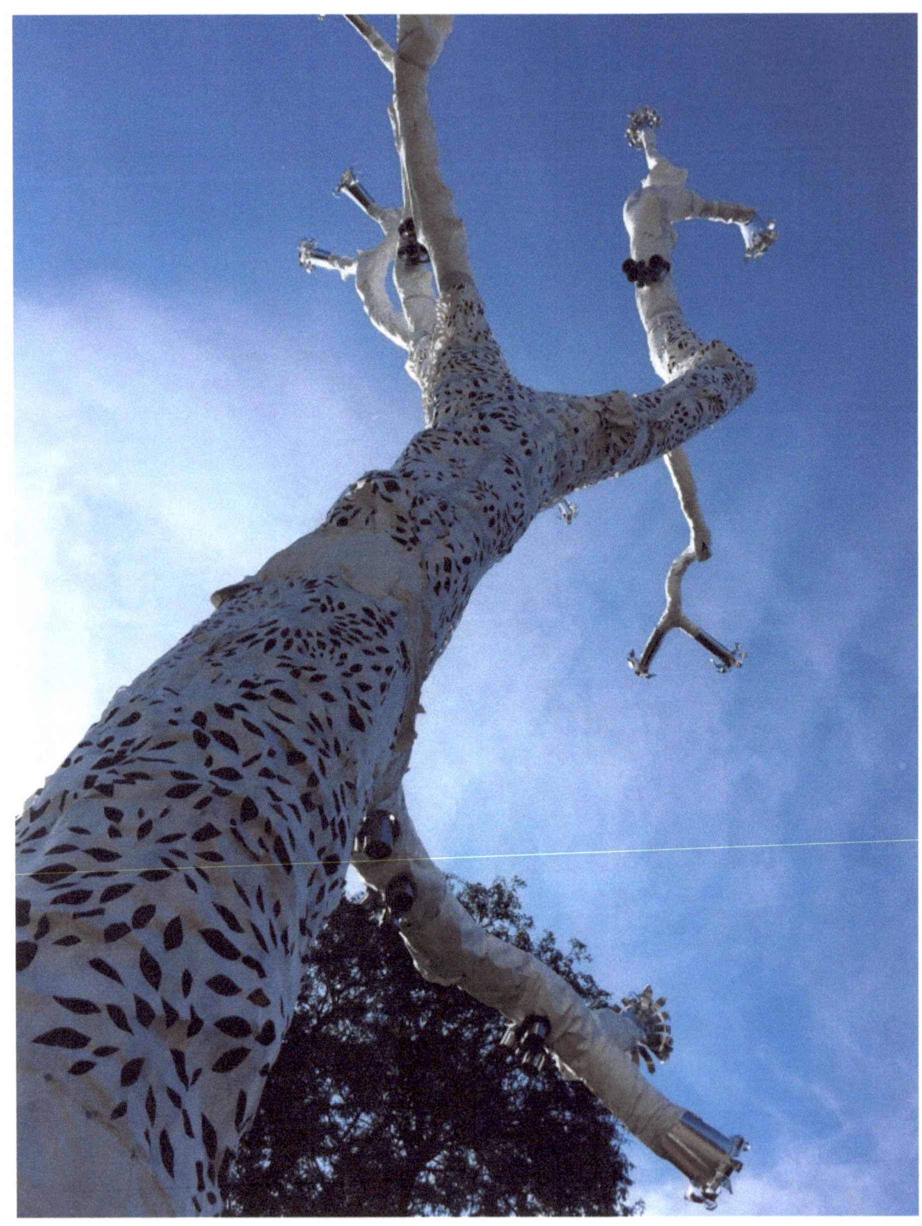

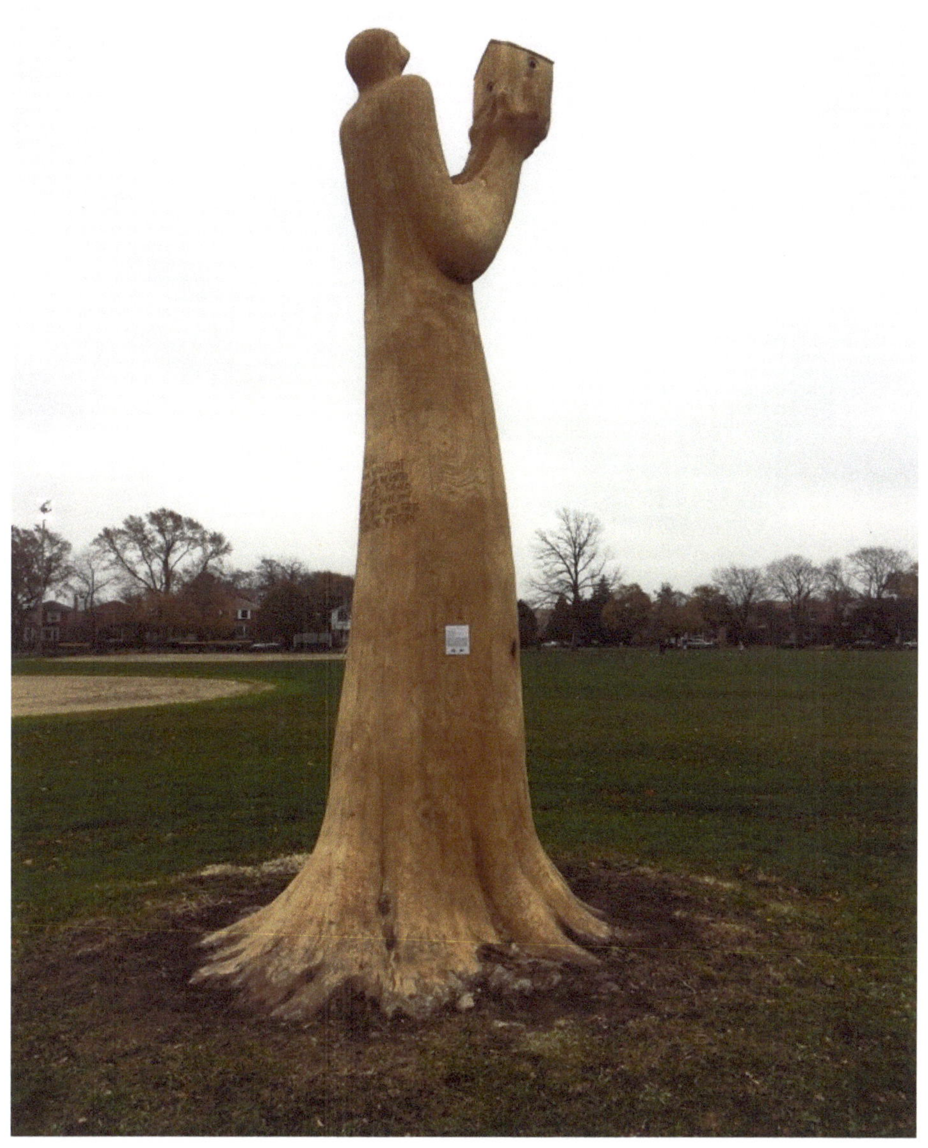

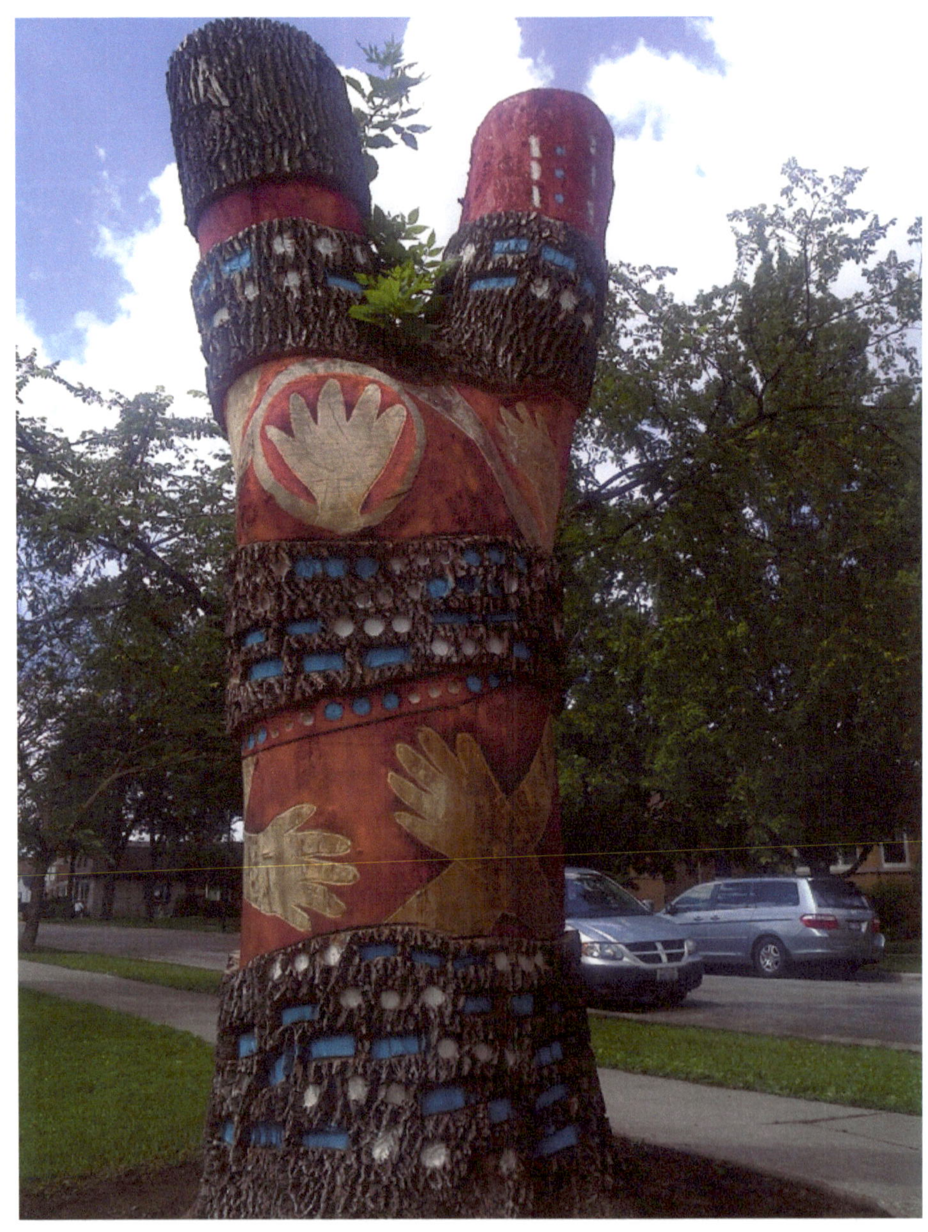

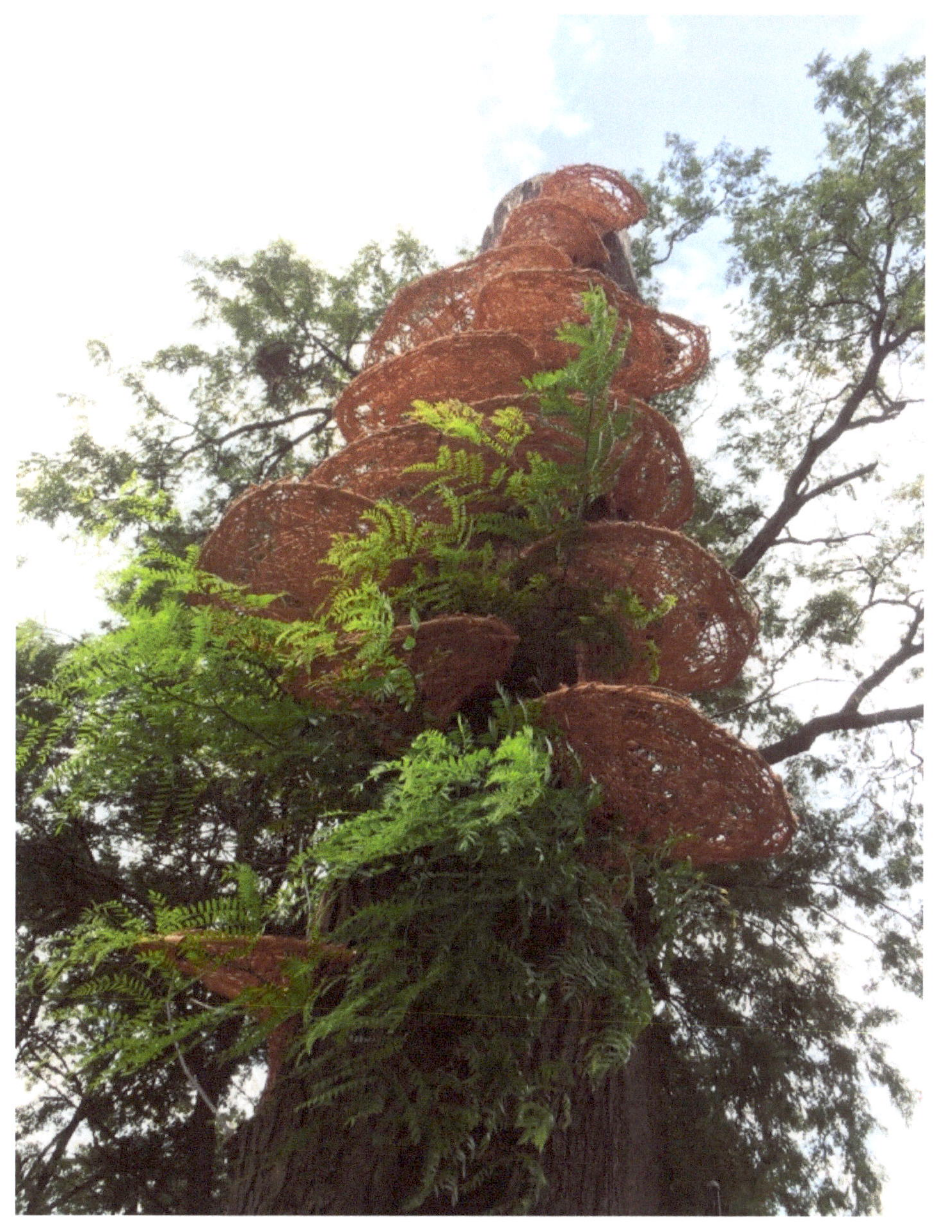

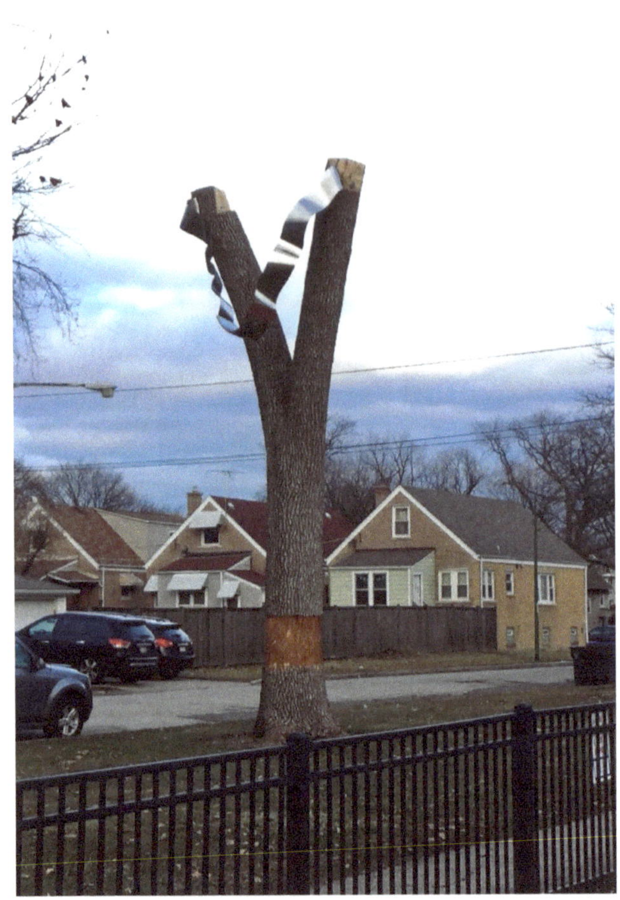

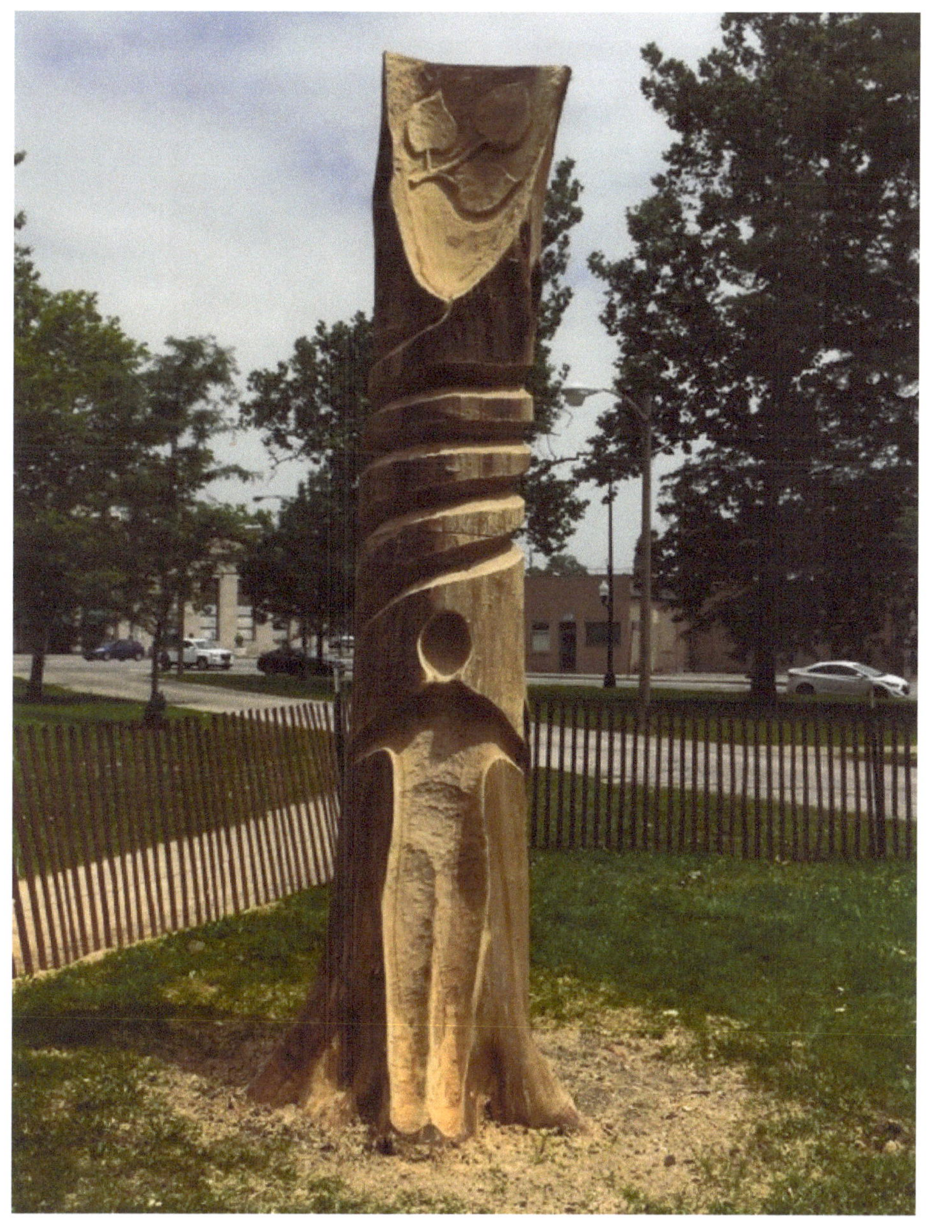

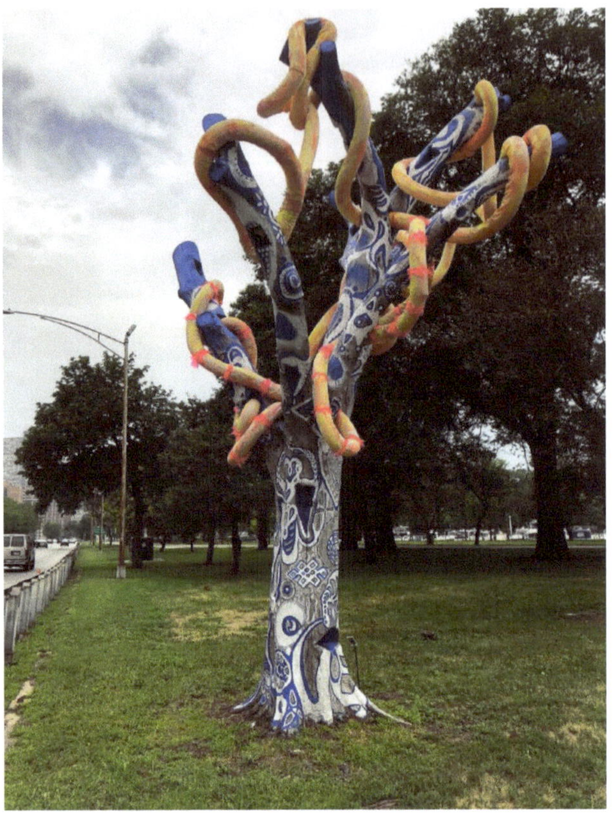

The Leaving Tree
My tiny nature sparked as a seed
Beneath a cap to become a tree.
My root wiggles down
Squiggles in the ground
Seeks water and fodder,
I shoot my stem to the sky, my leaves unravel,
Soaked with sun creates chlorophyll,
That feeds me to link my rhizome
To a forest family I call home.

My roots reason, reach to strive
Soak up nutrients for me to thrive.
My roots scheme their way around rocks
Winding in hummus, through earthly cracks,
Shoot tendrils 'round forest debris
Helping web with my mother's sharing,
 Her roots pumping food to my tender shoot.
 Slurp up puddles, brooks, and streams
 that are not too hot, nor too cold
Just right temperatures I handhold.

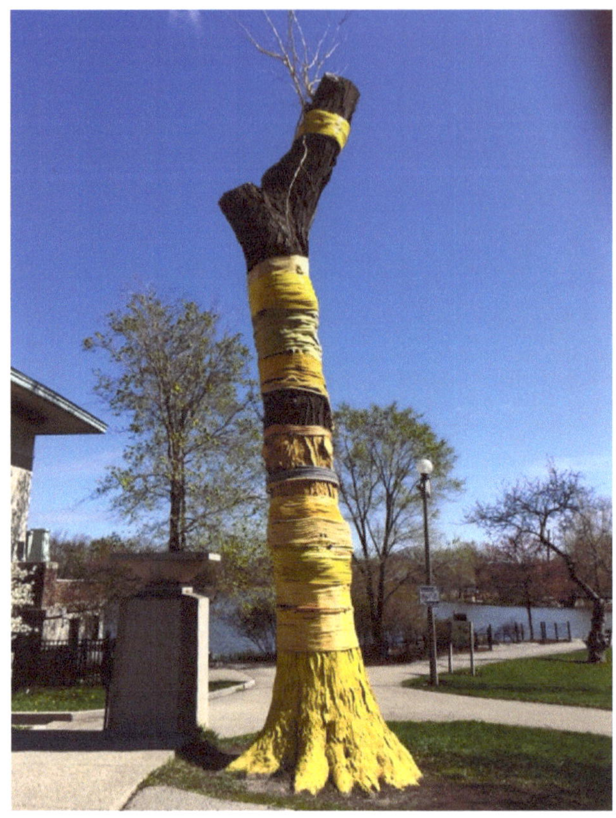

Connecting underground with like kin
Sucking oodles up my stem.
Orchard sprigs grow slow, wide and strong, as
Mother's canopy throws shade that is long.
We count spring days, eight, nine, ten,
Thirteen warm hours 'til days end,
Triggers my translucent bud-wrapped leaf
Curling, unfurling, twirling to greenery,
 Disentangle, dangle from my branch.

Reach to solar waves
Cupping falling rain
Shading, cooling, we scions,
Counting three, two one.
My leaves on broad branches soak seasonal sun.

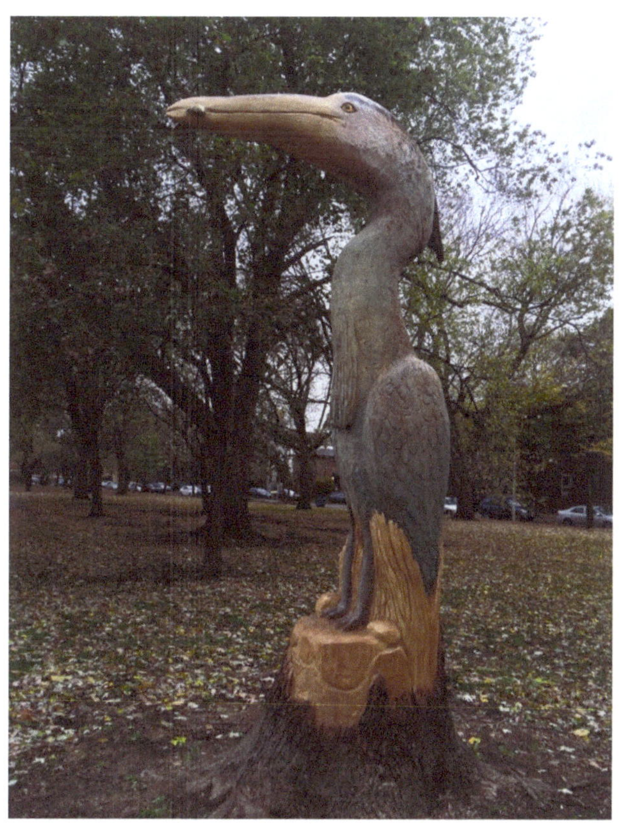

Decades circle my widen girth
From roots I stretch miles in earth.
Teaming with mycorrhizal furthers my realm.
Canopy shaded in summers, I rest while wintering
Waiting for my moment to burst forest's awning.
Successions, progressions during seasonal annuals
Grow strong and straight, my stalk blooms a parasol.
A river of water flows up my xylem
Sends chlorophyll production down my phloem
Swelling in rings each year through my cambium
Guards the white fan of mycelium,
I pull up water by the gallons.
To store in pores tons of carbons.

Plink, plunk seeds thunk my trunk.
Who knew that I could be
A tree to make a capped seed?
Splish, splash seeds do crash
Skittle, skatter round the tatter.
Chewed or dug in for delicacies
Acorns ensure descendencies.

Rat a tat tap tap
Red head rebounds back
From my smooth straight trunk
I look with disgust at a creature
That dares to invade my lacquer.
Beak breaks through my bark's defense
Ouch! I yelp to the offense.

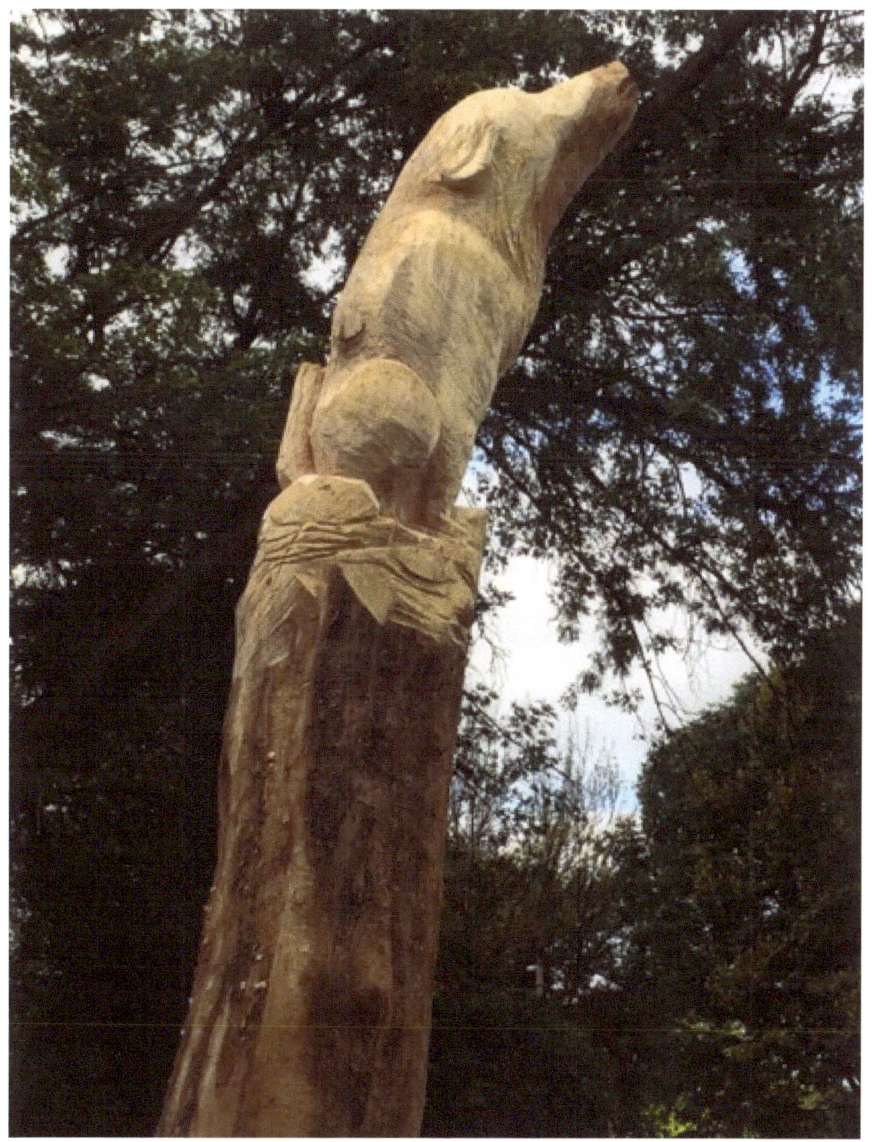

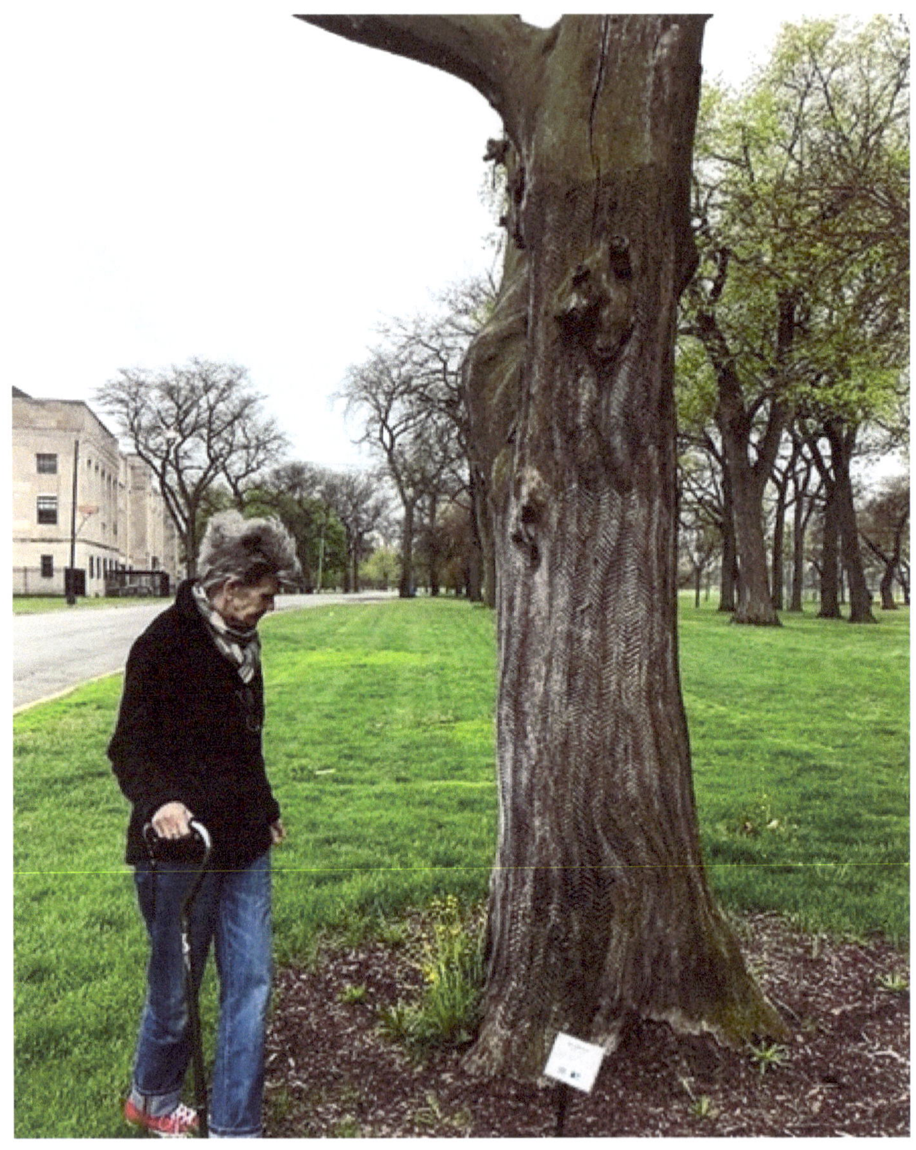

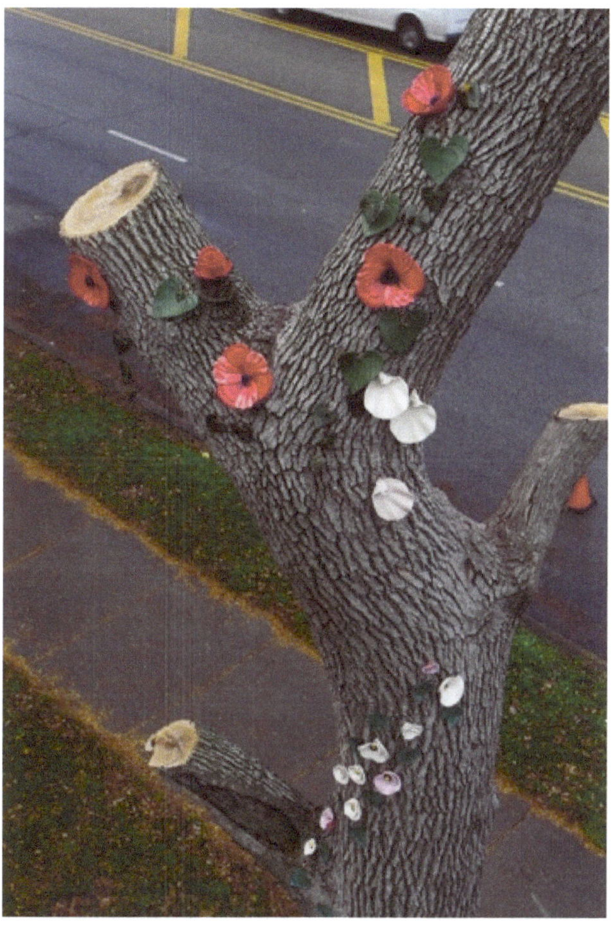

Scents are emitted as I react
To fungi creeping a silent attack.
Gentle slow fungus chews past
My excretions of scents-- toxin blasts!
I scream warnings to nearby trees
To excrete antibiotic barriers to these maladies.
Sharing aromas, odors, defensives

Declares of breaches and offensives.
Bacterial perfume fights off mold
Ensuring my family will grow old.
Warding off insects, aphids, and disease
Bellowing with distressed unease.
As my cambium's vertical flow breaches
By larvae in crannies, nooks and ruptures
Soft pulpy fiber breaks up strong wood
From the heart of my trunk that grew as it should.

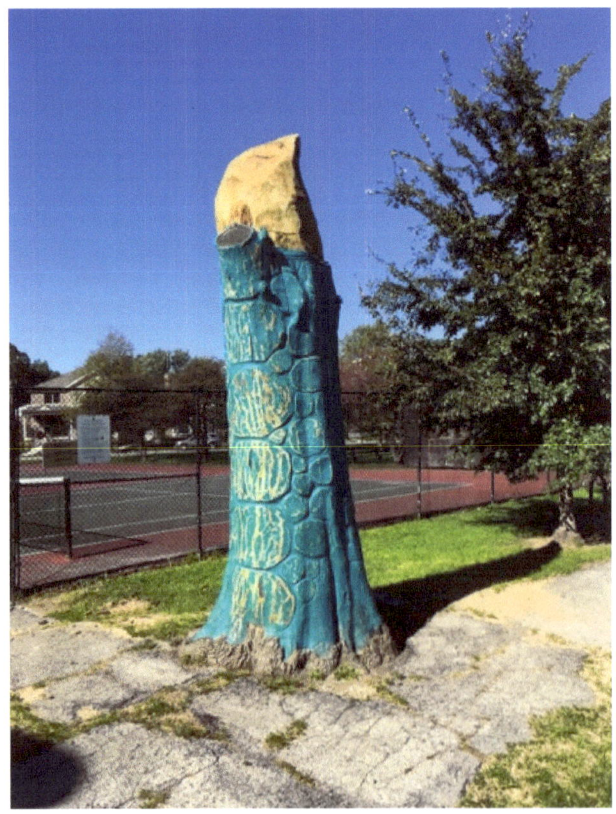

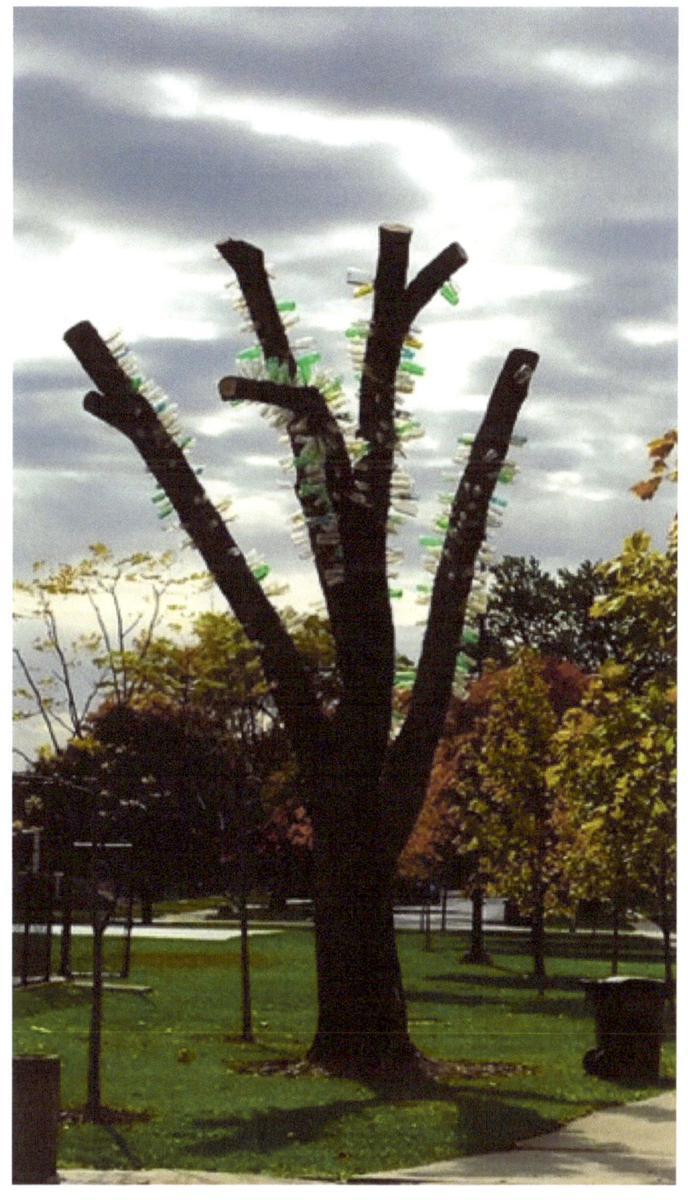

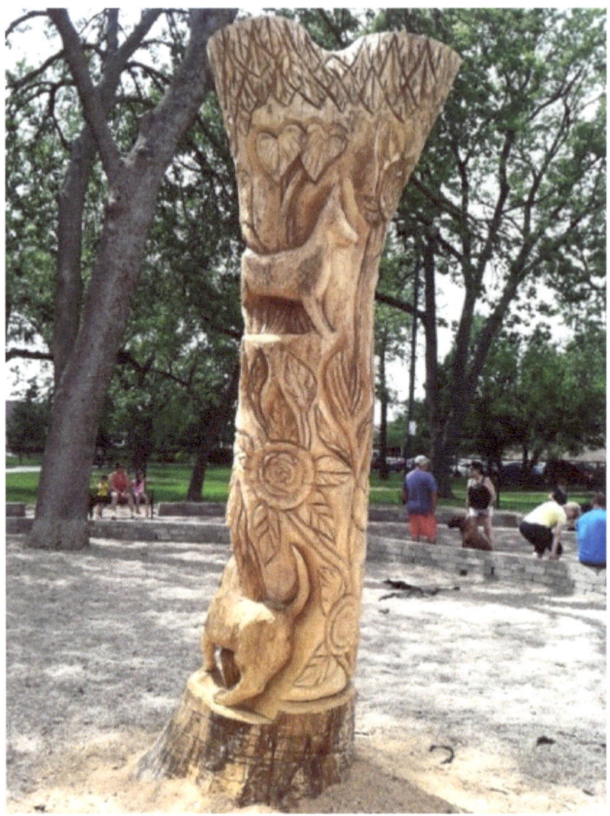

Cracks, fissures give in to invaders
Roly-polys, earwigs, bugs and arachnid
Dissolves the core of my heartwood
Opening wounds for intruders' food
To lay their eggs, make their home
Feed their young to flourish and roam.
My rustling leaves shake off interlopers
Gatecrashers, squatters, and imposters.

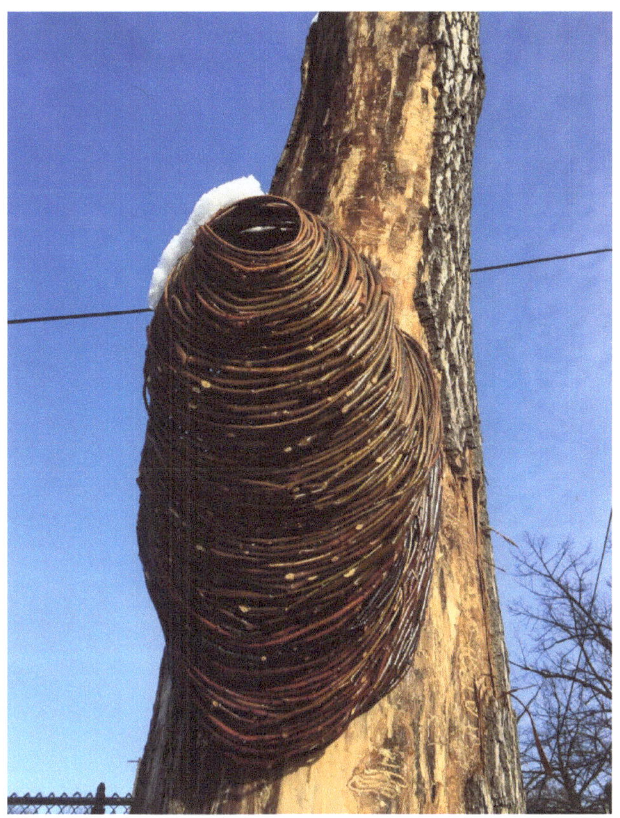

Rat ta tap tap!
Flickers back fast.
Pecking through soft wood
Flying off again.

Nestlings settle in enlarged cavities
Wrens, sparrows and chickadees
Varmints, small owls and mice
Fluff their nests within the raiders vice
Venoms emitted to the trespassers

My nutrients flood the door, sealing its passage
Growing a closure that's fed by tree's neighbors
All for one, one for all, we'll defend our corner
Share our scents, close our cupboard
We trees protect our hood.

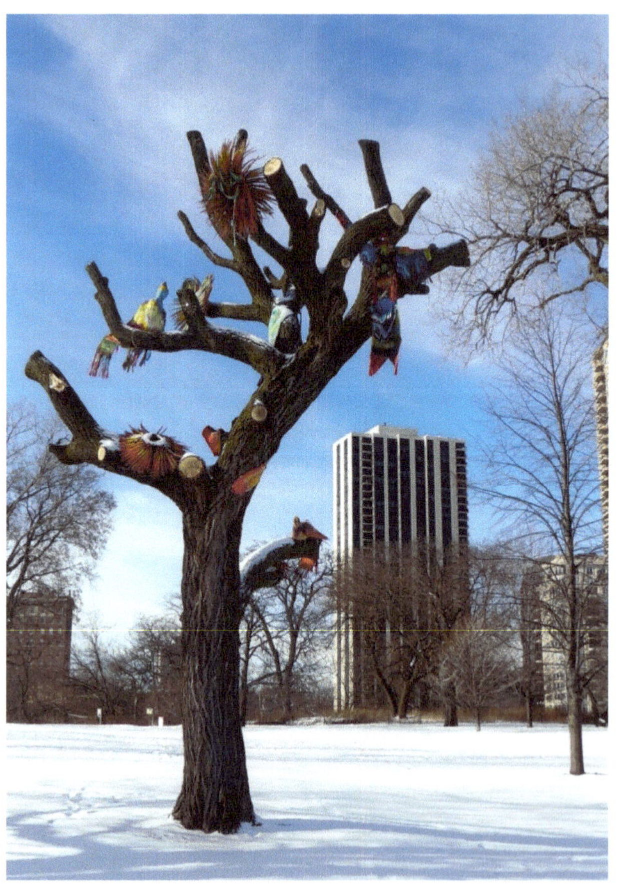

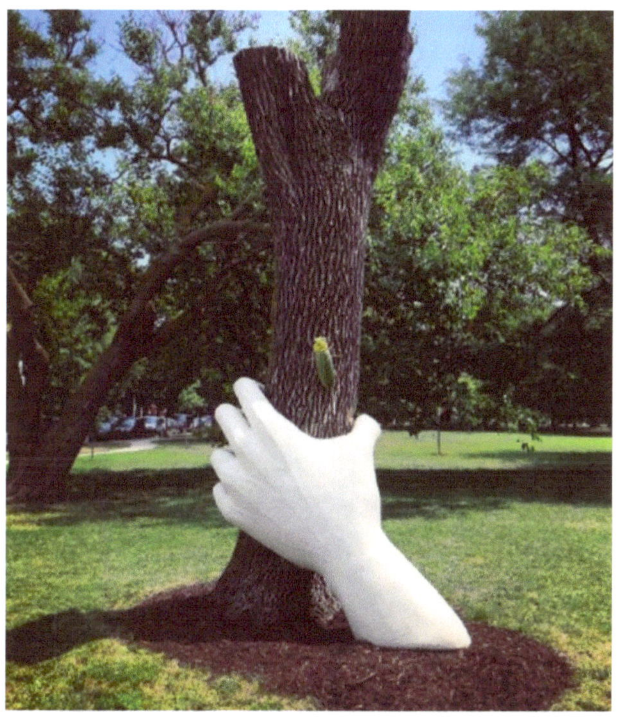

Together we show colors to signal warning;
Yellows, red, oranges, and purples are sporting.
Powerful leaves glow of health
Red states "stop there!" in vibration,
Orange utters "stand back" in celebration,
Yellow leaves speak of power and strength,
Purple proclaims to administrate.
Remembering as if we knew from all time
To count the cold nights; ones twos and threes
Before we begin to drop our leaves.
Twenty-seven square feet of thick compost
Around feet to feed underground roots.

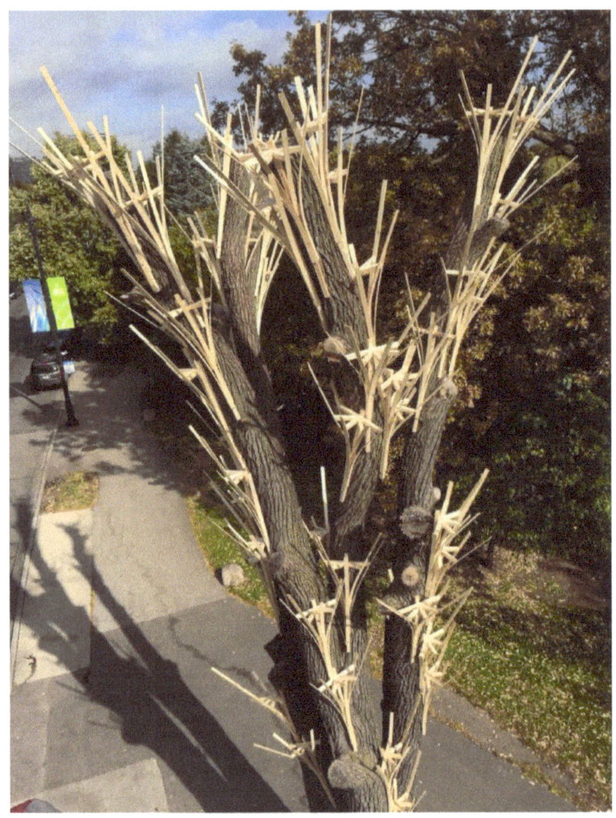

We offer you shelter, but don't go too far
Keep cambium's flow, and our bark from a mar.
Eat the heartwood if you must
For your future wealth and trust.
Keep your hatchlings growing strong
In our hollows teach no wrong.
"Leave my cambium be
To fuel my leaves," says me
So I give thee all I can spare
For the earth's creatures to care

For it offspring, its fledglings
For safe passage into Spring.

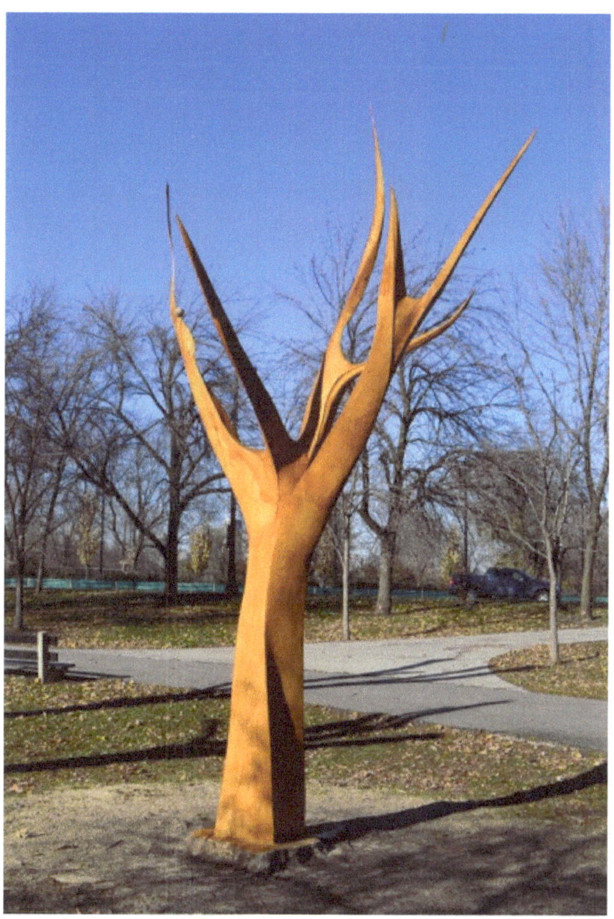

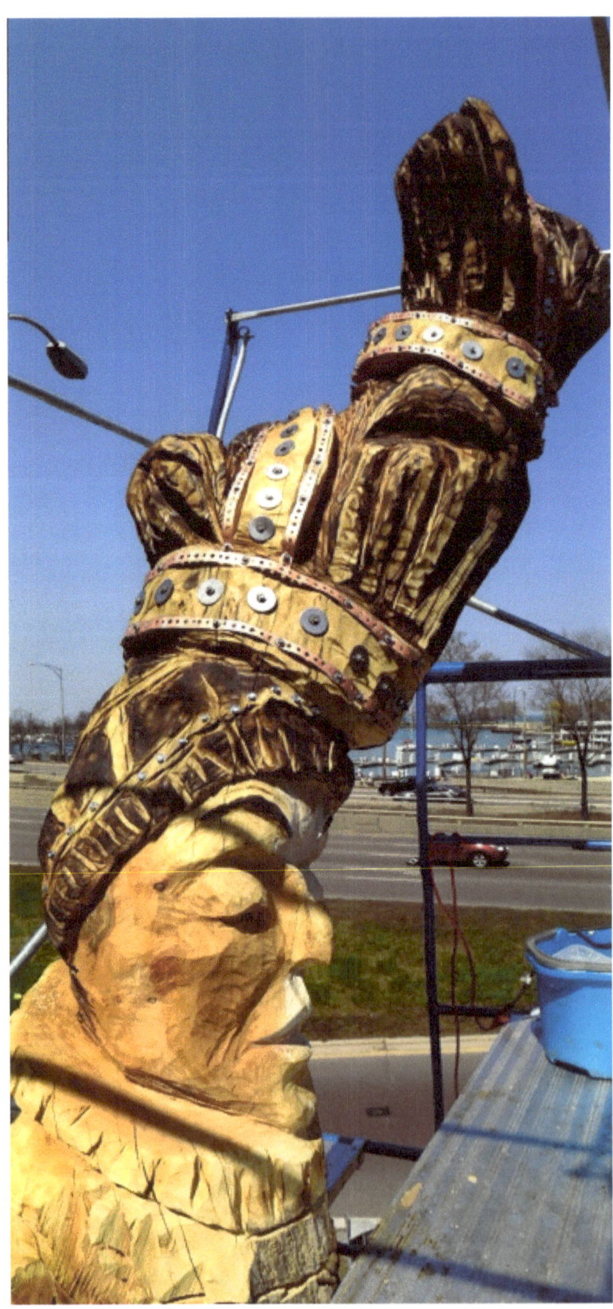

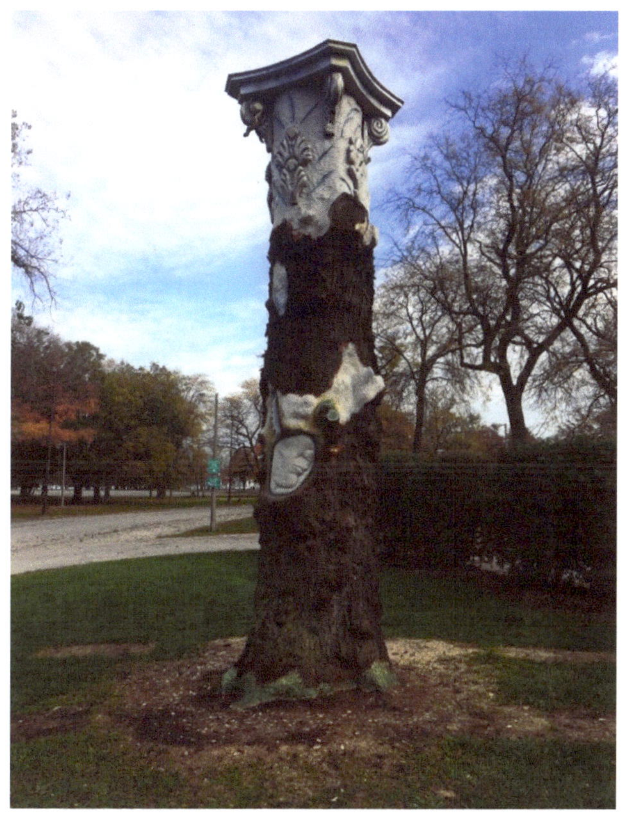

Rat a tat tap
Flickers come back
Rap to resonate my
Hollowed trunk reverberates
As my sound box attracts a mate.

Bats, birds, and squirrels, gather nests,
Claw, bite and spit my heartwood to chimneys,
Bees construct pockets for honey,
They conquer with swagger. They won!

To hover and protect their young.
Pirates of widening pockets devour
Clefts, splits and gaps
Into twaddle wood tower.

My burrowed branch plummets
On dropped leaves to decay to hummus.
Rain cascades through to soak the branch
 Breaking it down for roots' sustenance.Flooding the grove of saplings with sun
Flourishing, flowering, mushrooms my young.

A lumbering bear climbs to see
The honeyed hole within me, the tree
With long nails, she carves the core
Scraping out heartwood, bugs and more.
Occasionally she pauses to eat a spider,
Digs some more to fit her hide or,
Autumn's flesh to feed all winter.
While she curls amidst the splinters,
Cold subdues her hardly breath
Slows her heart to near death.
To feed slowly on her honey catch
Snuggling for frigid months of rest.
The early Spring sun angles in
Triggering her stomach's growling.
Finds her belly has swelled to
A woozy feeling of unwell.
She did thin to a silhouette
Slows her hunt for food to get.

Aches and pains,
her body strains.
Stretching joints from her curl
Waking her hunger, she unfurls.
She heaves and ho's to waking woes
As she creaks and crawls from the hole.
Two pups are birthed on spring's earth!
Two whippersnappers cling to her belly
As she hunts for bark and berry
To put an end to her hunger
As two pups ramble behind her lumber.

"Careful young sapling" tree warns
"Grow your stem strong, balance your umbrella
keep your roots cool, moist and healthy
allow for your consistency.
To live for many centuries"

Alone again I buffer rain,
Send my roots, grow a ring.
My green leaves unravel to gather sun
Turn to orange before autumns come.

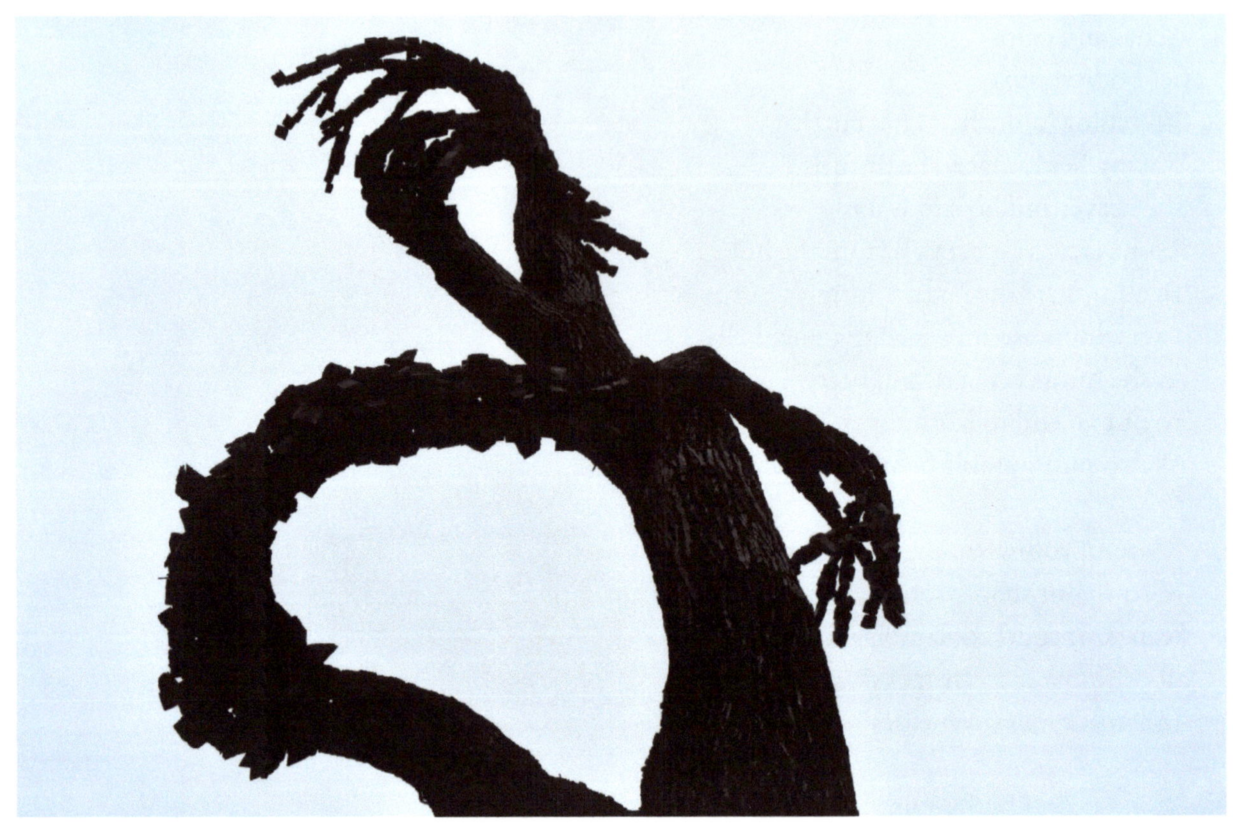

A ten-foot moose rubs mossy horn
Against my bark that is torn.
Lift its head to sniff the hole
Wonders if it will fit?
Who will feed a mother-tree
As I wither empty?
Strengthened by sharing

With the grove of trees caring.
Community feeds this hollow tree
Sharing sugars, water and company.

Careful young sapling, I warned.
Balance your parasol, grow your stem strong,
keep your roots cool, moist and healthy
allow for your consistency
to live for many centuries.

I grow faster than you, old tree.
The brash sun-soaked youth replied with gaiety.
Gobbling up carbon, spewing oxygen
Chasing the brightest rays of the sun.
Gulping up water with abandon.
Bursting with energy of the woody decay
Collapse from the hierarchy.

I store tons of carbon extracted from gusty breeze
Slowing the wind with my expansive canopy,
Steadying my neighbor to recoop from the breeze,
Calming the clime for your comfort and ease.
Take the fallow of my branch for feed.
Grow slow little tree," replied me.

-----Margot McMahon

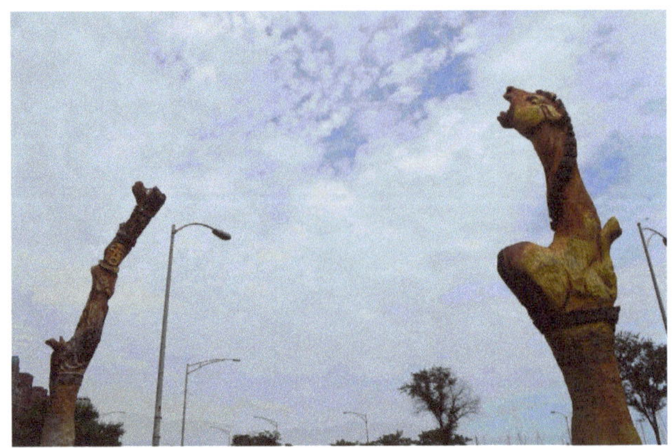

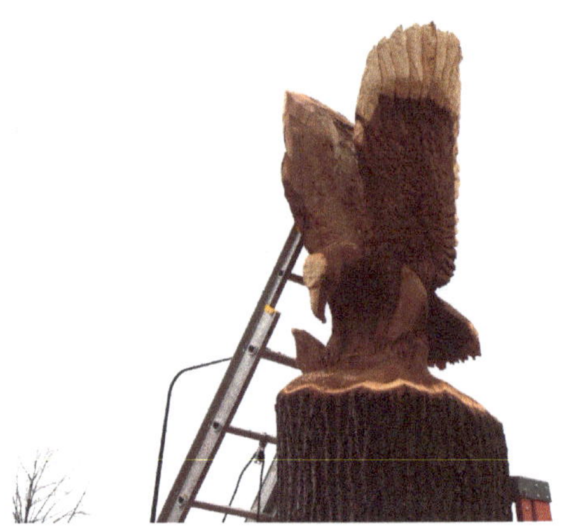

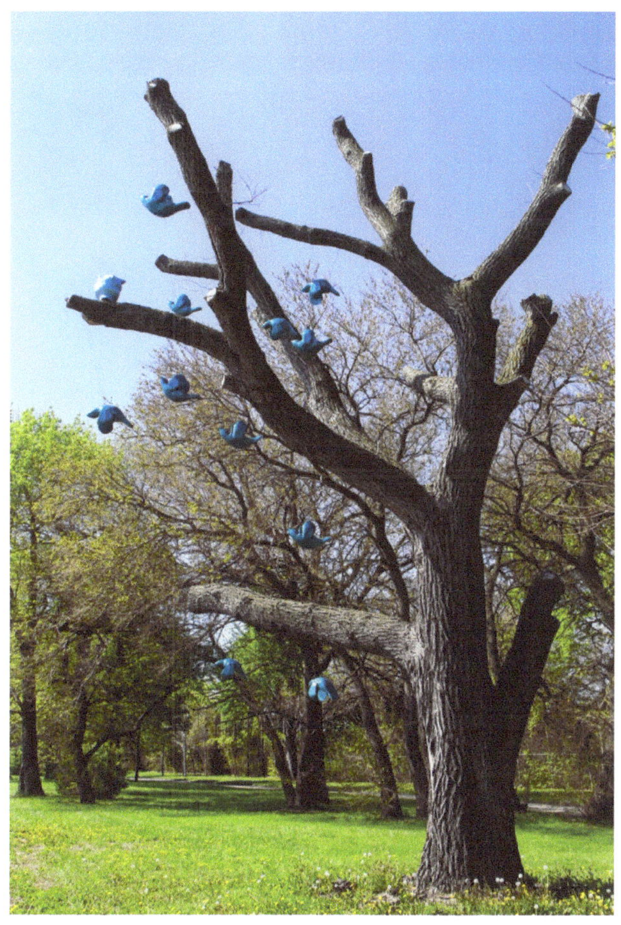

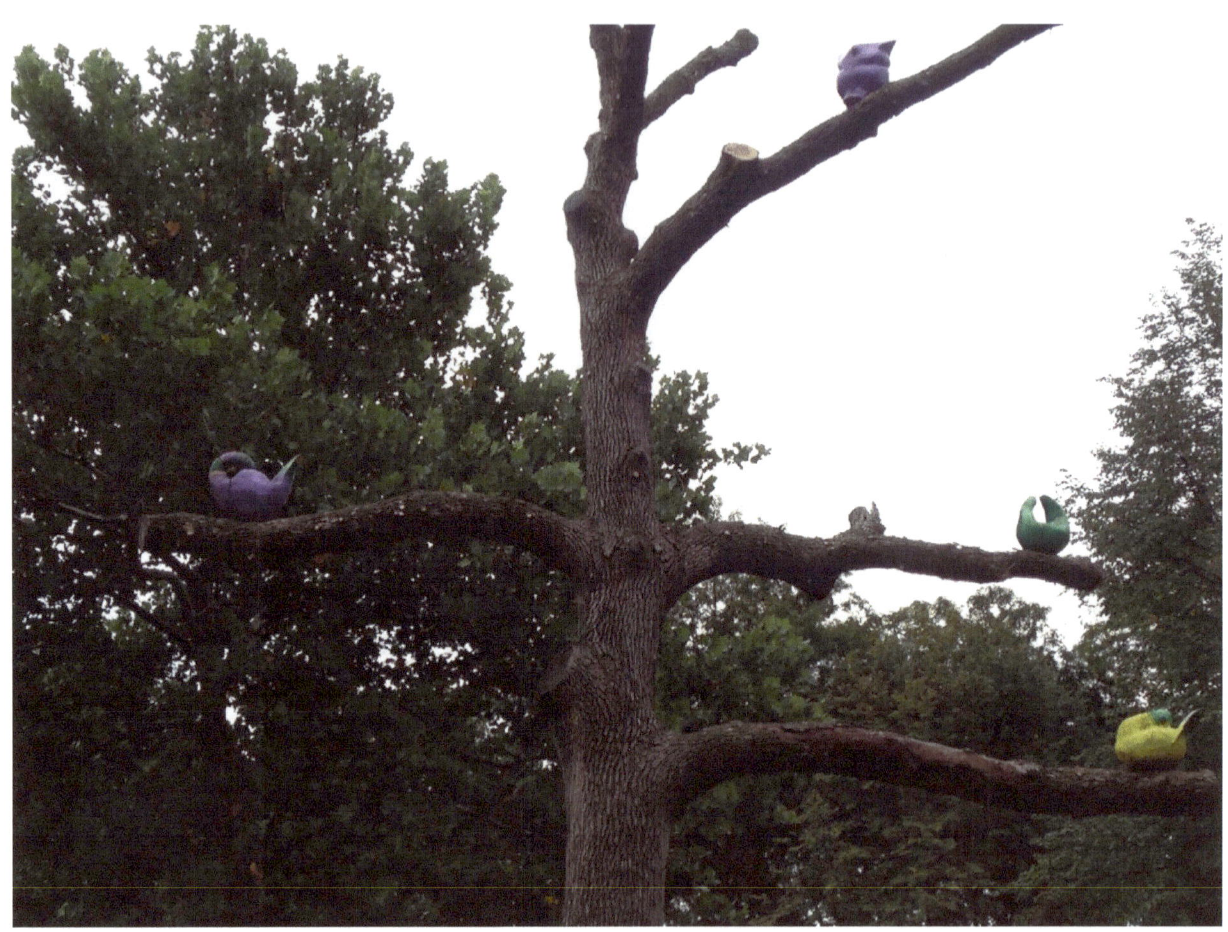

Chicago Tree Project Trolley Tours

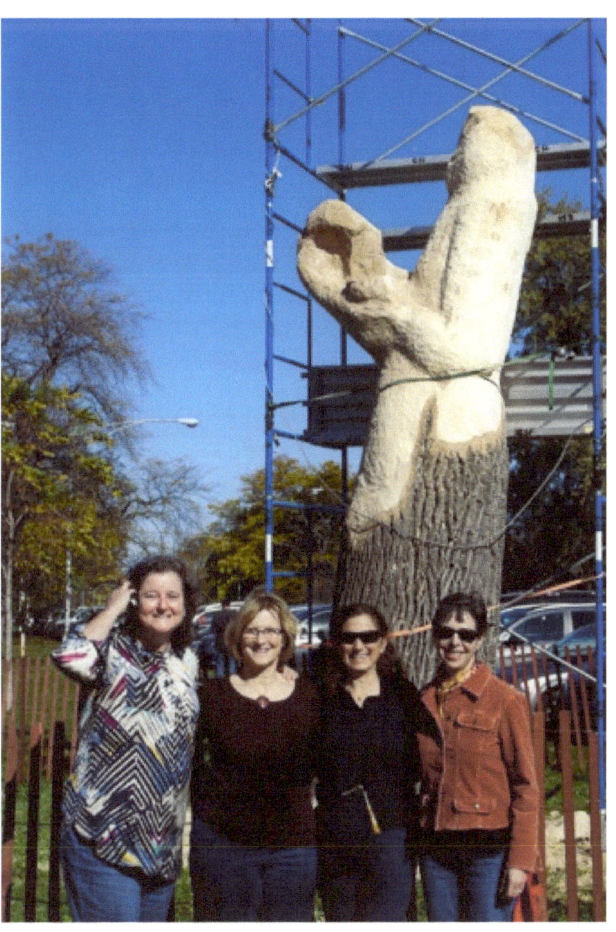

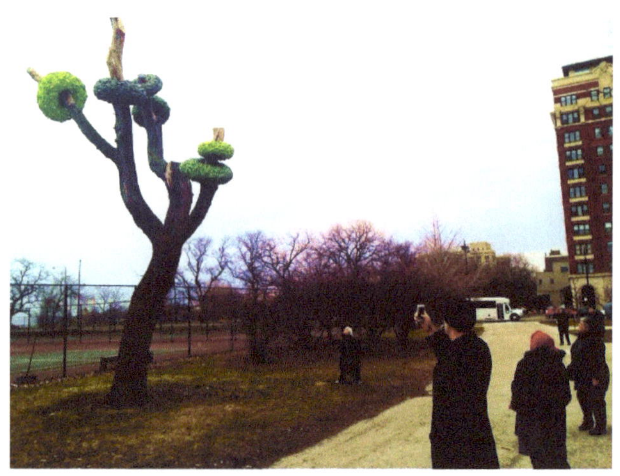

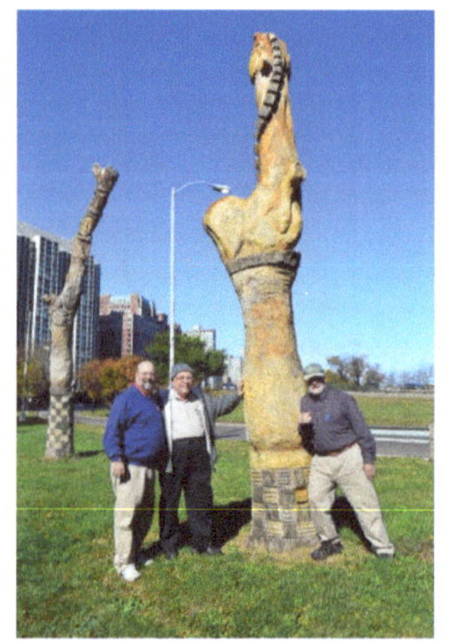

ASAP! Adopt a Sapling Project

It's easy to become a tree steward! Basic activities include watering trees, adding mulch and soil, and removing weeds and litter; as well as advanced activities such as installing a tree guard, expanding tree beds, and installing or removing stone or brick pavers.

Basic Tree Care Activities

Tree Care Tips: Watering

For the first few years, watering is the most important thing you can do for your tree. It may also be the hardest task to accomplish. Getting water from the source to the tree can be a challenge. Also, because of pollution and compression due to foot traffic, city soil has difficulty absorbing water. This means that you need to cultivate or loosen the soil so that the water can reach the tree's roots. There are a number of different tools and techniques available to help your watering efforts.

Tools You Can Use

- 5-gallon bucket
- hand cultivator
- hose
- Treegator & Treegator Jr.
- hydrant adaptor
- trash-can

Suggestions for Street Tree Watering

- Poke small holes at the bottom of a large trash can. Fill it with 15-20 gallons of water and leave the trash can next to the tree overnight.
- Ask building maintenance staff to water trees while they are hosing off sidewalks.
- Ask street vendors and merchants to dump water from their containers (coolers with melted ice or flower buckets) into nearby tree pits at the end of the day.

• Make sure water with detergent or bleach is dumped into the gutter, not the tree pit.

Weeding your site, although not as important as watering, is still essential. Both street trees and green streets live in very small spaces that provide limited amounts of soil and nutrients. Weeds are fast-growing and fast-reproducing plants that tend to dominate space and sap resources from the plants intended for the site. If weeds are left untended, they will ultimately kill some plants and stress others. Thus, weeds should be removed from green streets and street tree pits as frequently as possible.

Tools You Can Use

- gloves
- trowels
- garbage bags

Guidelines

- Identify the plant as a weed. Visit Rutgers University's Weed Gallery for help in weed identification.
- Wear gloves. When removing weeds, take out the entire root system. Leaving behind some of the plant will allow the weed to grow back. Use trowels or weeders to dig out stubborn roots.
- Put the plant and its roots into a garbage bag, or compost it.
- Dispose of trash properly.
- Do not use a weed whacker or lawn mower next to a tree. They can damage the bark and weaken a tree.
- Reduce herbicide use near a tree and in surrounding lawn.
- When weeding, be on the lookout for poison ivy and any hazardous trash in the soil.
- Try to remove weeds before they flower and spread seeds.

- Weeds are easier to pull when the ground is wet. If the ground is dry and the weeds are stubborn, soak the ground with water first.
- If you find white wood aster growing under your tree, consider keeping it rather than pulling it. It is a native wildflower that will eventually form an attractive ring around the tree. Plus, it will produce pretty white flowers for several weeks each fall.

Waste

Keeping a site free of litter not only ensures its place as a community asset, it reduces the amount of stress placed on the plant.

Guidelines

- Keep dogs and dog waste (both liquid and solid) away from the tree. The waste will overwhelm a tree, burning its trunk and throwing soil nutrients out of balance. Encourage dog owners to clean up any droppings within the tree bed (The City requires property owners to clean up animal waste on their property and adjoining sidewalks and gutters even if they do not own the pet, stray, or wild animal making the waste).
- Keep garbage and de-icing salt out of the tree pit. Try alternatives to rock salt (sodium chloride) such as calcium chloride, granular urea, sand, or sawdust. In the spring, flush the tree pit with water to dilute winter salt buildup.
- Pick up any discarded cigarette butts within tree pits. The inorganic material in cigarette filters will not decompose and the chemicals in the cigarette can damage the tree.
- Politely inform other members of the community about the damages caused by allowing trash or animal waste in tree pits. Some residents may not know that their actions have a negative impact on the tree's survival. Dogs are staples of many NYC communities and they are even more appreciated when handled by responsible owners!
- Consider installing a tree guard or signage for your tree to discourage people and animals from using it as a garbage receptacle.
- Tree Care Tips: Planting Flowers and Shrubs

Planting around street trees and yard trees is recommended if done carefully, however, aggressive flowers and shrubs compete with the tree for limited resources. Find out more about what to plant – and what to avoid – within a tree bed.

Guidelines for Planting Vegetation in Street Tree Pits

- Perennials, annuals and bulbs are beautiful additions to a tree pit, as long as you remember that the tree's health comes first.
- Some perennials can be complementary, but flowers that have shallow roots and die back each year (annuals) are recommended.
- Choose plants that require little watering. Key words to look for are "drought tolerant" and "xeric conditions".
- Use small plants and bulbs – large plants require large planting holes, which damage tree roots. In addition plants with large root systems compete with the tree for water and nutrients.
- Please do not plant flowers within 1-foot of the tree trunk.
- Do not add more then 2" of soil to your tree pit. Raising the soil level will harm the tree.
- Mulching a tree pit is always good for your tree and plants. Mulch keeps the soil moist and prevents weeds from sprouting in tree pits.
- Be sure to provide enough water for the tree, not just enough to perk up the flowers.
- Greenstreets are individually crafted by our landscape designers and maintained by gardeners; please do not add extra plantings to them.
- Never plant bamboo, ivy, vines, woody shrubs, or evergreens. They are all major competitors for water and nutrients and can stunt or kill a tree.

Tree Care Tips: Mulch, Compost, and Soil Maintenance

Mulch is roughly ground up wood from other trees, and can act as both a protective buffer and nutrient-rich layer for a young street tree. Spreading a 3-inch layer of mulch on a street tree bed can help prevent soil from becoming hard and compacted, making it easier for the tree's roots to take in air and water. Mulch is also a great retainer of moisture, and can help prevent the soil from becoming extremely dry during a hot summer drought or high winter wind. In addition, spreading a 1-inch layer of compost, decomposed food, and yard clippings before placing the mulch will allow the soil to better hold air and water, drain more efficiently, and provide a nutrient reserve that the tree can feed on.

Too much mulch, compost, or building a mulch 'volcano' around the base of the trunk can be harmful for the tree. Before any additions to the tree bed are made, it's important to aerate the soil. See how it's done below!

Tools You Can Use

- Gloves
- Cultivator
- Rake
- Bucket or Wheelbarrow

Guidelines

- To aerate the soil of your tree bed, take your hand cultivator and rough up the dirt 1-inch to 3-inches down. This will break up the compacted soil and allow more oxygen to get down to the roots.
- You can obtain mulch through stewardship other street tree care workshops; or by becoming a Care Captain. You can buy mulch at gardening stores like Home Depot, and sometimes mulch is available for stewards in NYC's parks.
- If you do not compost at your home or workplace, you can obtain compost from your local park district
- Spread compost to cover the whole street tree bed. The layer should be no more than 2 inches high, and should not be touching the trunk of the tree. You should be able to fit your fist between the compost and the trunk.

- Spread mulch to cover the whole street tree bed. The layer should be no more than 3 inches and should not be touching the trunk of the tree. You should be able to fit your fist between the compost and the trunk.
- If you do not have enough mulch to cover the whole bed, build a ring of mulch 3 inches away from the tree trunk. You should be able to see the roots at the base of the tree taper off into the soil.
- Do not build a mulch volcano. A 'mulch volcano' is what happens when mulch is piled high and close to the tree trunk. This does not allow the roots and lower part of the trunk to interact with the atmosphere and encourages 'girdling,' a root problem that eventually causes the tree to strangle itself. Mulch volcanoes also create rot by building up too much moisture against the tree trunk.
- Although mulch can be applied at any time of year, spring and fall are best for young trees to help retain moisture and prepare for extreme seasonal temperatures.

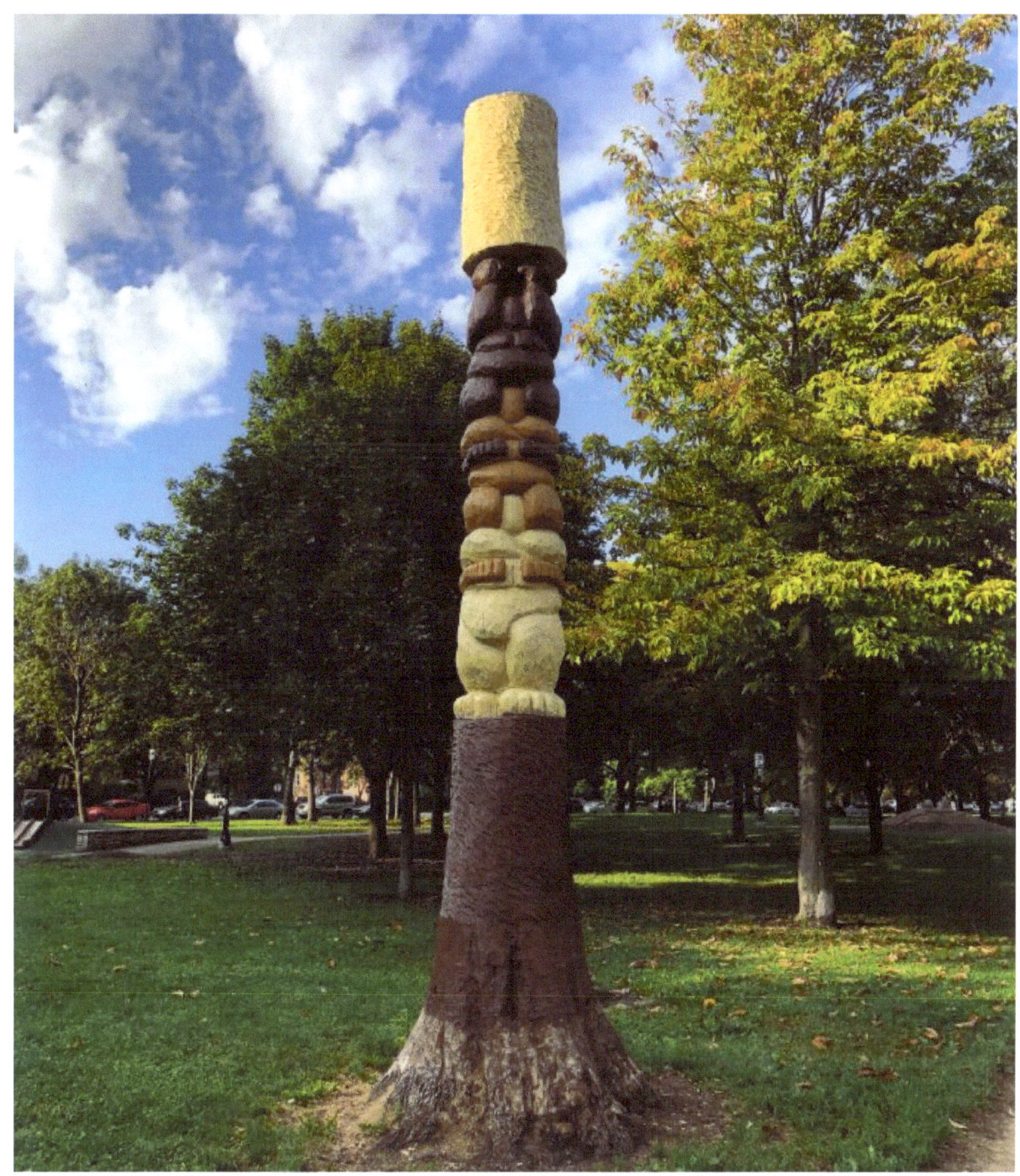

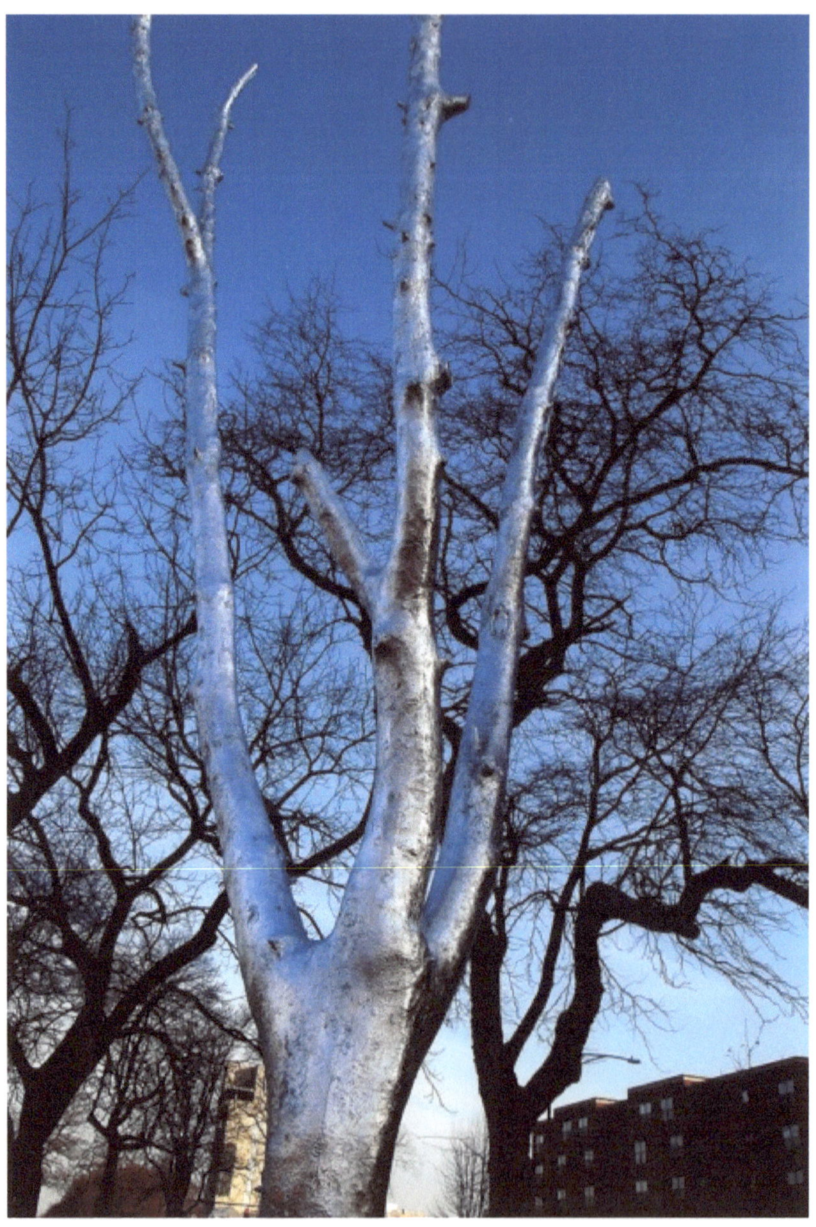

Chicago Tree Project Artists

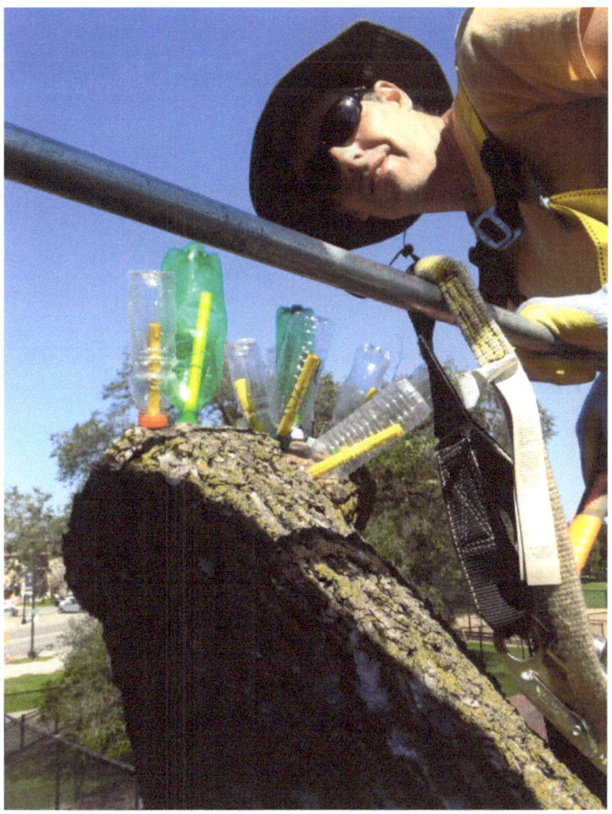

Tess Landon statement

Since 2014, through a partnership between the Chicago Park District and Chicago Sculpture International, artist from across the region and the country have come to Chicago to transform dead trees into works of public art. Since the appearance of the Emerald Ash Borer beetle in the United States in the early 2000s, tens of thousands of trees in Chicago's parks have been killed or endangered by this invasive species and the Chicago Park District has worked quickly to take down infected trees and replace them with new varieties. In order to bring awareness to this environmental issue and to beautify our shared spaces, the Chicago Tree Project turns a small portion of these trees into temporary artworks. The trees serve as a reminder of the fragility of our earth and bring the community together both during their creation and throughout their newly extended lifespans.

Christine Perri - "De-Constructed Monument"
Vivian Visser - "Ledge Fungus"
Anthony Heinz May - "Progress > (Product)", "Natura Non Confundenda Est"
Allison Svoboda - "Ghost Tree"
Kara James - "Save the Birds - Thank You" "Lead with the Heart"
Margot McMahon "Perch • Preen", "Flock", "Checkmate: "Queen" and "Knight"-Lincoln Park
Thom Cicchelli - "Arbor Renewal"
Samantha Rausch - "Endless"
Gary Lehman - "Shards of the Rainbow"
Indira and Karl Johnson - "SOS, Question Not Answered"
Nicolette Ross - "Woven"
Ron Gard - "Dying to Survive #3, #6, #9"
J. Taylor Wallace - "Aspire"
Marc Schneider - "Invasive Species"
Mia Capodilupo - "Burst!"
Karen Gubitz - "Second Chance" "Gaggle"
Vivian Visser - "Hive"
Phillip Shore - "Eternal Connections"
Jim Long - "Howlin' Wolf" "Ol Heron" "Fishing Eagle"
JR Cadawas - "Dog Tree"
Janet Austin - "The A'maze'ing Larvae of the Emerald Ash Borer"
Janet Austin and Emily Moorhead-Wallace "Nestfull""
Actual Size Artworks - "Conundrum"
Sandra Bacon & John Hatlestad - "Fanciful Birds"

Nick Goettling - "Untitled"
Tracy Ostmann Haschke - "Fraxinus Raptor"
Elain O'Sullivan -"Slingshot #1"
Connie Noyes - "The Sound of Light on Ash"
Samatha Rausch - "Endlesss"
Steve Gaeth - "Recompense"
Peter Krsko – "Para Z"
Carrie Fischer – "Helping Hand"
Irene Hoppenberg "Lemon Tree"
Gary Keenan "Transformation"
Peter Krsko "Para Z"
Johnathan Schork "Thalidomide #12"
Berthold Boone "Readymade"

Please visit chicagotreeproject.org for more information.
AudioTour: https://www.otocast.com/destinations

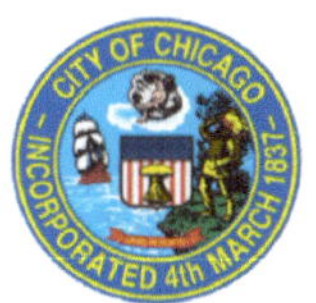
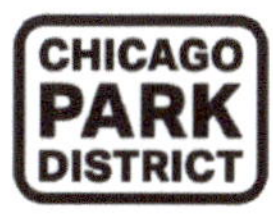
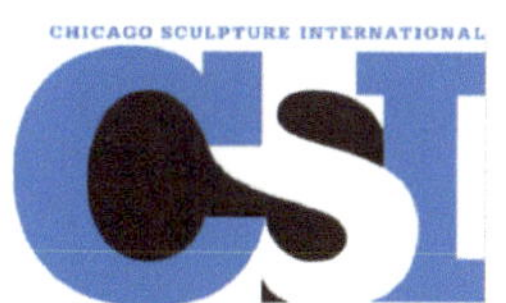

www.ingramcontent.com/pod-product-compliance
Lightning Source LLC
Chambersburg PA
CBHW042114030726

47599CB00002B/210